HUMANITY'S DESCENT

Other Books by
Rick Potts

EARLY HOMINID ACTIVITIES AT OLDUVAI

RICK POTTS

Illustrations by Jennifer Clark

AVON BOOKS
A division of
The Hearst Corporation
1350 Avenue of the Americas
New York, New York 10019

Frontispiece by Lynn Sures
Published by arrangement with the author
Visit our website at **http://AvonBooks.com**
Library of Congress Catalog Card Number: 96-1699
ISBN: 0-380-71523-6

Published in hardcover by William Morrow and Company, Inc.; for information address Permissions Department, William Morrow and Company, Inc., 1350 Avenue of the Americas, New York, New York 10019.

The William Morrow edition contains the following Catalogin in Publication Data:

Potts, Richard, 1953–
Humanity's descent : the consequences of ecological instability / Rick Potts ; illustrations by Jennifer Clark.
p. cm.
Includes index.
1. Human evolution. 2. Human ecology. 3. Natural selection. 4. Biotic communities. I. Title.
GN281.4.P65 1996 96-1699
573.2—dc20 CIP

First Avon Books Trade Printing: May 1997

Printed in the U.S.A.

QP 10 9 8 7 6 5 4 3 2 1

CONTENTS

CHAPTER I

ORIGIN

I

ARCHED over a sun-seared land, the afternoon sky swirls with dust and dismal clouds. The cliff where the sun sets, crowned with thorned acacias, stretches out above the canyons. But now it cannot be seen. In a remote gully carved out of an African plain, our singular collection of black and tanned bodies seems unmoved by this strange event. Sixteen men and women keep looking down, seeking pieces of bone hardened by the ages and stones fashioned by an ancient hand. Our attention appears to center on the parched ground. The thud and slow rasp of metal tools on hard sediment speak of disinterment, broken pieces of animals, buried signs of an ancestor, clues of ancient environs. Yet the mind's eye and occasional glance heed the slanting curtain of torrent veiling the near horizon. An hour ago, the group of us huddled in this anthropologist's pit was beaten by the sun, sweating, needing water. Now we are about to be drenched by the sacred substance.

The weather will end our day's search. Reluctantly, we will cram ourselves into a mechanized metal box, protected from the elements. The truck will take us to our hearth on the edge of a cliff a mile away. This happened yesterday and the day before. It is supposedly the middle of the dry season, the period we crave as excavators searching on the edge of midnight in the heat of an equatorial, rift-valley day.

Wrapped in red-and-white-checkered cloth, the local elders tell us they do not know why the rains are coming now. It is late August; October is the month the dry season ends. After another two days of sudden torrents, they will explain "A four-day rain in late August is not so uncommon." Their memories serve them correctly. The dryness will return. Our temporary lives here will proceed as the wind, dust, and heat allow.

Much of my time is spent digging in the rift valley of East Africa. My immediate goal is to exhume fossil traces of early human lives and environments; but this is only the start of a curious path. All too often in this line of work, the central effort—or *hope*—is to find the crown jewels of human origin, fossilized skeletons like the famed "Lucy," or the bony fragments of nameless ancestors. Not in this case. Our team strives to see ancient worlds—places inhabited by unfamiliar antelopes, short-necked giraffes, odd and diverse elephants, and a bipedal ape with the dawn of humanity in its eye and a tool of stone in its grasp. They have all gone extinct.

The emergence of humans to dominance on the world ecological scene has happened over a very long time. It is a fragmented 5-million-year episode encased in sediment. The ecological history of our kind is not easy to perceive in its vast complexity. Rocks that hide the clues cannot be read like newspapers. Fossils do not speak. And so our work—excavators, geologists, and lab workers of diverse backgrounds—appears to uncover only the most distant glimmers, knowledge held far away in the grasp of an unfamiliar past. Yet these glimmers illuminate where we stand and how the kiln of evolution forged our apparent dominion over nature.

We are a foul-weather species: Humans do not just withstand change; they adjust to it and make their own modifications. We enclose ourselves for protection and ride out almost any storm. Animals are the subject of conservation; for them the issue is how not to become extinct. *Homo sapiens* is somehow above that. We both *cause* extinction and motivate ourselves to prevent it. We build storehouses of knowledge about the environment to know what happened when. Our need for control over nature spawns mental databanks about monsoons, blizzards, and other recurrent events that affect our ability to reap, and occasionally even to survive. This is the fruit of human culture, a harvester of world resources for its purposes.

If you've read the newspapers recently or seen the sections of bookstores devoted to Environmentalia, you know that astounding troubles threaten Earth's atmosphere, oceans, and land. We eliminate forests, thicken the air with exhausts, reduce wild populations to extinction, modify the flow of water, and change the world climate. Our ability to alter the state of the planet is a global dilemma that highlights the supposed conflict between man and nature.

The burden of human activities appears to confirm our separation from nature and affirm our special origin. It is now our responsibility to try to be good managers of the planet. The irony is that this image is born of the same "we-can-do-it-because-we're-different" creed that has fueled our powerful exploitation of the natural world. The ecological problems caused by human wastes, overpopulation, and habitat decimation *and* the idea that people are the planet's managers are two sides of the same graven image: human superiority and separation from nature.

Our inquiry here offers a different slant on humanity's genesis, a novel way of thinking spurred by discoveries about ancient habitats and our ancestral ways. The fossil record yields the solid residues of actions carried out by early humans. A stone tool, for instance. The oldest known tools are not just rocks chipped and sharpened on purpose. They also tell us of pieces of the environment obtained over 2 million years ago and held in the hand of a being who was not quite us. Considering a multitude of clues, we can trace the development from a way of life once shared by the common ancestor of humans and African apes to another way: human dominance on Earth.

By placing the fragmented events of prehistory in their original settings, we find that human life was boldly transformed by its intimate relationship with the natural world. Human forebears emerged on a stage of climatic decay, followed by dramatic swings in weather and in plant and animal life. Earth's environment has not been a stable status quo, now disturbed by or protected from human intervention. Rather, our activities are closely, endurably connected by millions of years with the shifting environments of the planet. Our present activities and ecological burdens mirror the environmental conditions in which we evolved.

It is important for us to meet those ancestors who lived in settings so different from today's. These relatives, from a biological standpoint, are dead. They flourished well before our own time. Individually, we are so ill prepared to deal with the deaths of people in our personal lives that it seems like very distant stuff to be concerned about the deaths of species akin to us a very long time ago. But the birth of humankind would have been impossible had they not lived. The meaning behind the extinction of these family members is worthy of our attention.

From an ecological standpoint, we can ask, "Who were these relatives?" If we could reverse our geologic telescope and have *them* look at *us,* would they feel proud of their descendants? Or awed by the landscapes paupered of species that we now occupy, and by our own critical contributions to this state of affairs? Did they become extinct only to give way to an ecological bully who will bring world ecosystems to a devastating crash? Exactly how different is the world we live in now from that of our progenitors?

In pursuing these roots we need the vantage point afforded by millions of years. Climate watchers typically deal in decades, very rarely in a century of change. Other observers of the global environment search the last ten thousand years for trends. But there is a much longer perspective still to learn about.

My own excursion into the environmental legacy of our kind starts in the ground. Places with names like Olduvai, Zhoukoudian, Kanam, Sangiran, Lainyamok, Koobi Fora—the strange language of Stone Age sites. It begins wherever the adumbration cast by sediments on our ancestral path can be parted, slowly, exposing with each unearthing season the conditions under which you and I were born from earlier experiments in uprightness, tool use, and enlarged brains. This is something we can investigate. We can ask questions, propose answers, express doubts to one another until new approaches and clues refine the answers. Chemicals encased in old soils and bits of fossil animal are signals of a hidden past and there is a tide of exploration that can read life, landscapes, migrations, and survival, millions of years old, from exposed pieces that are now as lifeless as a rock, or are, in fact, rocks.

In the rift valley of southern Kenya, a place named Olorgesailie first sparked in my imagination the excursion in thinking that we will take. The southern rift is inhabited by the pastoral Maasai. Indeed, Olorgesailie is a Maasai word, enunciated Oh-lor-geh-sigh-lee. This is the name of a mountain nearby. It means "the place of the Giselik people," though no one knows who they were.

Our small coterie of excavators and scientists has traced the geologic lines of old landscapes, the rhythms of change, the systolic spasms of lava from beneath the ground. And this journey, not yet complete, has led me to delve into the sequences of climate and conditions of existence in other places where human forebears, the hominids, have lived over the past several million years.

Because it is in the bottom of the Great Rift Valley, Olorgesailie is an open sanctuary of prehistory, a land of white dust where, in daytime, the twirling wind picks up the blanched skeletons of microscopic algae, called diatoms, and brandishes them high in narrow funnels called dust devils. The ubiquitous white hills and plains of Olorgesailie are tremendous piles of these desiccated diatoms, which once thrived in a lake now gone. Ever since I first laid eyes on Olorgesailie, in 1977, the idea of a large body of water turned to white dust has struck me as mysterious, a beguiling irony of nature. The spiral storms of dust are a reminder throughout the day, as though a whisper of inordinate influence rides on nature's passing breath. By nightfall, the dust devils are replaced by a cool, continuous wind, stirring the eroded gullies and the taut canvases of our tented camp.

An escarpment adorned by a tall, gangly tree is the center of our home here for a few months each year. It overlooks an interwoven series of slopes, rich in fossils and the residues of old makers of stone tools. All around us are breadth and age. The contours of erosion and lines of contact between sediment layers are like the passage of time in a great-grandmother's face. These slopes are worthy of respect.

The dimension of time lies rugged on the landscape. Passing visitors usually do not see it fully. It needs a knack for seeing maturity in a wrinkle of the land. Dormant volcanoes dimly record a youthful, angry gesture filled with fire and magma. Now, covered by moss and grass, their eroded sides are graceful signs of aging. The dipping troughs in the distance are filled with old cinders and other chronometric debris drawn from these erupted highlands. A kind of chemical ticking in the sediments testifies to a time when one of the volcanoes erupted: 1 million, 100,000 years ago . . . and something's happening. Though the fact is usually overlooked, Earth has long been in turmoil.

Olorgesailie is one of many places in which to discover how ancient versions of humanity lived, how they made tools, carved a joint, smashed a limb bone. But an unanswered question remains: By what process did these simple acts and the survival of these predecessors give way to the present ecological handiwork of the human species?

The most important journey we must take is mysteriously connected with, yet very distant from, the present. It is to a period alien to us, the times of our ancestors. It is strange to us because it seems so long ago and has been made obscure and unrecognizable by the many revisions of evolution.

Many similes have attempted to capture the immense time recorded in the tree of life. Among the most common, the standard twenty-four-hour day stands for all of Earth's history, contrasted by the brief moments of human ancestry. The earliest humans, according to one version I've read, are poised on the edge of midnight, near the end of the twenty-four hours, just before the chime strikes the arrival of the present.

I like that image—searching on the edge of midnight for clues to our origin. But time offers an amazing tension between short and long. The brief minute or two before midnight that encompasses all of human evolution actually entails 5 million years. This amount of time is a dull blur against daily schedules or the concreteness of a lifetime. But we must bring that blur into focus if we are to learn the process by which humans arose on Earth.

The ecological history relevant to human origin goes back even farther—50 million years ago, when Earth began to undergo a serious decline in temperature and a deterioration in environment that, oddly enough, would prove to be our evolutionary fortune, the setting for the birth of our lineage. This turn of global events was also vital to the vegetation and many animals living on land today. It is by patient and intricate detective work that this unfamiliar setting long ago may become familiar.

2

Virtually all attempts to explain how humans originated begin on common ground. They seek to define the special qualities of the human species in relation to other animals—distinctions in behavior, social life, and anatomical structure. The special and extraordinary features of humankind then become the building blocks of our origin—our intelligence, technology, ability to speak, body posture, dexterity, ability to learn, social complexity and ability to create governments, marriages, alliances, and other institutions.

What specialists in the field of human evolution have long tried to explain, however, is not really the development of intelligence, or toolmaking, or language, but humanity's superiority. All of the elements are seen as interdependent, so tightly knit in the emergent fabric of our origin as to be inseparable. To explain the emergence of any one element, such as braininess and acumen, you need go only as far as one of the other elements in our story of success—the importance of tools, or our unique social behaviors. Orthodox renderings of human origin look no farther than the internal properties of our species.

I find it intriguing that the greater the gap conceived to exist between people and apes in any particular trait, the greater the role that trait is given in the emergence of humankind. In his attempt to determine the root of our lineage, Darwin surmised that the use of tools and the tendency of people to occupy open terrain rather than forests were the two fundamental differences between us and our closest animal relatives.

Drawing on the work of his close friend Thomas Henry Huxley, Darwin stressed that the African apes—the chimpanzees and gorillas—are the closest relatives of mankind. This meant two things: first, as Darwin proposed, that the earliest humans also evolved in Africa; second, that to comprehend the continuities between human life and the animal world, we must use the African apes as the vital point of departure and comparison. It is not your obedient dog or your clever cat, not the bees who communicate by a shimmying body language, not the intelligent dolphin, nor all the small-brained approaches to life that abound around us. All such comparisons between *Homo sapiens* and these other animals are of passing interest, inappropriate from an evolutionary standpoint, which will always tend to confirm the separation between humans and other species, preserving the special place of our kind in creation. Darwin and Huxley's breakthrough accented the continuities between life-forms. It is with the apes that humans find their connection with the tree of life.

By a detailed comparison of anatomy, Huxley showed that the gap between humans and apes was no greater, sometimes even smaller, than the distinction between apes and monkeys—between a chimpanzee and a baboon. In the shape of the rib cage, the loss of a tail, the character of their teeth, humans and the African apes share unique features that separate them from other primates. And if there is a hiatus in the structure of the brain, and not just in its size, it interrupts the continuity of life between the extant monkeys and lemurs rather than between humans and apes.

These observations were made long before there was any knowledge of DNA, the genetic code of organisms. After its discovery, it was clear that this molecule at the center of all cells records the biological relationships between species, the branching and connecting paths among living beings. It is the comparison of DNA that confirms what Huxley and Darwin urged. You and I share a significantly greater amount of our genetic code with the African apes than they (or we) share with the baboons and other monkeys. Between humankind and the apes, the gap in anatomy, bodily functions, and genetics is less than it is between apes and other animals. The molecules of biology make this an indisputable fact.

In Huxley and Darwin's time, a more shocking contradiction could hardly be cast on the assumptions of Western thinking. Some quarters of our society, entrenched and protected by bunkers of belief in the eternal separateness, superiority, and dominion of our kind in nature, still reel from these findings.

In the 1870s it was completely unknown that some of the apes could make crude tools, that certain populations inhabited dry open areas, and that all lived in organized social groups. Because these things were unknown, toolmaking, terrestriality, and social life composed the vital distinctions, the foundation on which Darwin's scenario of human emergence was built.

The late 1800s saw growing acceptance of Darwin's argument that evolution connects all species in an enormous genealogical tree. Darwin even dared to explore the implication for human beings, showing that the unique and universal characteristics of humanity could also be explained as a transformation linking us with the rest of the animal world. His task was to cover all the bases, bringing each distinction that was surmised to exist between the African apes and humans into the theater of human descent.

In 1871, Darwin surmised that human ancestry began with a grassland ape in Africa. As a continuous band of rain forest fractured and widened into treeless terrain, the threats and opportunities of living on the ground required a means of protection, and since the human lineage lacks the large eyeteeth possessed by apes and other primates, the development of tools for defense was the original, key adaptation of protohuman life on the savanna. Dependence upon hand-held tools created the conditions for walking upright, for brain enlargement, for cunning, culture, and the forging of new social contracts. The human condition emerged in a swirling interplay between new behaviors and anatomical possibilities.

The first, swift kick was provided by the savanna, which required its apelike inhabitants to survive and produce young in a new setting. This process—the accepting-and-rejecting impact of environmental change—was the crux of Darwin's vision of *how* evolution occurred, the theory of natural selection. The primal accommodations of the human lineage were developed in these apelike populations, and once this happened, the unique array of human characteristics inevitably followed. The evolution of one trait stimulated the emergence of the next in a chain reaction.

Since Darwin's time, the science of paleoanthropology has debated the ways in which the pieces of human uniqueness best fit together. Two-legged walking that freed the hands, technology and fine manipulation of tools, hunting and eating meat, large brains, culture and language, the social complex of the pair bond, division of labor, and marriage, all developed in tandem as a complexly interwoven package. Because these features are bound together in modern people, it is easy to assume that the package must have emerged within a brief interval. If evolved beings we must be, the characteristics defining our kind had to evolve together and are therefore solidly integrated. In applying this approach to human origin, any of the key differences—toolmaking, hunting, sexuality, or sociality—could be nominated as *the* prime mover, for all the rest would follow like a cascade after passing the first rough rocks in this new and untried river of evolution.

By the mid-1960s, Jane Goodall's startling observations that chimpanzees make crude tools out of sticks, leaves, and stones had dramatically diminished the presumed gap between apes and humans in regard to toolmaking. Now we know that certain populations of chimps in

West Africa even create something resembling the oldest archeological sites—hammerstones and collections of food debris, in this case nutshells broken by using the stones. All of the apes, in fact, manipulate objects in the wild with fine dexterity and surprising insight.

Over the past thirty years, the complexity of ape and monkey societies has been amply documented. As a result, one of the most dominating preconceptions of the last century—that humans possess organized societies, while animals have only the very rudiments of social organization—has been proved wrong. While important ape-human differences exist in social life, the actual formation of societies in which individuals possess recognizable roles can no longer be seen as either a unique product or a directing cause of human origin.

In *African Genesis* (1961), Robert Ardrey introduced to the world at large the hypothesis that hunting and a murderous life-style played the leading role in the origin of the human lineage. As a dramatist and convert to the study of animal behavior, Ardrey's aim was to magnify the views of the South African anatomist Professor Raymond Dart. From several thousand fossilized bones of baboons and antelopes, associated with the then oldest known hominid fossils, which were about 3 million years old, Dart figured that the worst had happened—our kind had adopted the survival track of the carnivore and evolved the aggressive impulses of a killer ape. Dart's purpose, and also Ardrey's, was to elucidate the annihilations and executions promulgated by the human species over the centuries. Both proclaimed that warfare and hunting were unknown in the gentle apes, and thus the mark of these behaviors upon the human lineage furnished the axis around which the human endeavor evolved.

Here is another example where the key unlocking the mystery of human origin is the element of human life deemed the farthest removed from the lives of other animals. However, hunting, meat eating, and the realities of intergroup aggression are now known in our closest biological relatives, the chimpanzees; cooperation in hunting is also seen in more distant kin, savanna baboons. These observations in no way mean that hunting and meat eating had no influence on the emergence of the human lineage. Indeed, in all of these traits—how we get food, our enormous social complexity, our impressive technology—the differences between humans and nonhuman primates never cease to amaze. Human uniqueness remains intact. But the gap, that all-important separation of humans from beasts, is smaller than was believed possible. By exploring the animal kingdom, we have found that the continuities are far more dramatic than students of evolution ever expected. The differences are of scale and degree, not of kind. The popular concept of *the* missing link, once deemed the Holy Grail of my field, has disintegrated. The bridge between humans and other animals cannot be captured in a single fossil find or archeological site. Instead, a multitude of links over time, involving anatomy, behavior, society, and ecology, now enlivens the research on our origin.

In focusing on large brains, technology, culture, and other unique aspects of human life, the orthodox study of our origin has turned these special features into success stories of how humans came into being. It is the emergence of human dominance in the animal kingdom that has grabbed our attention, and all that is needed to account for it are the defining attributes of modern humanity. To explain the rise of technology, mainstream anthropology points to the importance of culture and, farther back in time, to the advantages of new food resources, such as meat from large animals, which were possible to acquire with tools. Enlarged brains are explained, according to one theory, as the result of making better tools, or of sharing food, and other innovations of social life, or—still another theory—of an emerging ability to speak using symbolic language. These are hailed as the ways humans have overcome the obstacles that limit other organisms, the ways in which we have achieved our dominion.

My attempt here is to cast human origin in a different light, a viewpoint initiated by Darwin: The evolution of animals and plants is guided by the conditions in which these organisms live, conditions created by climate and physical habitat, other species, and members of one's own kind. The revolution in investigating human origin revolves around the ecological genesis of *Homo sapiens* and its ancestors, by which I mean the emergence of human qualities as a response to natural conditions.

The major events and developments in human origin were not inevitable improvements for the success and ultimate dominion of humanity; they were part of Earth's ecological past, consistent with its environmental history. Humanity evolved in a halting manner as environments became less predictable and more varied from place to place. Deterioration and change in habitats were hallmarks of nature well before our species emerged as a significant ecological factor. The two-legged toolmakers who survived were those able to cope with fitful alterations of their habitat.

The central principle of our evolutionary response is flexibility, the ability to adjust and diversify our behavior, physiology, and overall way of life. In the face of an erratic habitat, no better coping mechanism exists than the ability to modify one's surroundings. The ability to alter is, however, itself a product of nature, of the environments in which human ancestors lived, and the pace of change in these settings. In the end, we—the survivor—have acquired a ponderous capacity to alter our surroundings and, therefore, to mimic the very processes of environmental change that helped to create us. Our ecological characteristics were honed by vicissitude, not by the static measure of nature we typically adore.

In short, our ecological capabilities mirror the pace and degree of nature's continual face lift. Much as we may want to praise the intrinsic aspects of our origin—the presumed success of *Homo sapiens*—human evolution has been an ecological genesis, a response to factors external to us, namely, the exigencies of Earth's history.

Ours is a species bent on writing its memoirs. Our origin and history anchor us. And because we write with an eye on the eras that will become history, the biography we tell garners more than mere sympathetic judgment or curiosity. It is the beginning of all that we want life to be, or must grow to accept it to be. The topic of origin ultimately concerns the reasons for human existence.

The problem is that no topic could be more prone to the present agendas and often mythical influences that serve particular interests and creeds—economic, political, moral, or religious. People formulate the way they think about the world in their accounts of its origin. They relate their idea of interaction with the world, with nature and with other human beings, in these same accounts. Origin is a bold statement of relationship and expectation.

From the late 1800s until well into this century, virtually every scenario of human evolution presented a single line of progress leading inexorably to modern man. The history of life that European and American authorities recounted was a matter of advancement beyond "the inferior," the success and dominion of human beings. It has been pointed out that this progressive view of origin sharply mirrored prevailing economic pursuits and outlooks on social progress and destiny.

On closer inspection, there is a pernicious element in certain ideas about origin. Into the mid-1900s, the prevalent views of evolutionary human advancement were deemed to show racial progress that, in effect,

1870 – 1970

defined the exploited peoples of the world as unequals. Like creation stories around the globe, the Euro-American perspective produced only what it could, a tale that rendered the narrowly defined perspectives of its people.

Other prejudices were also infused into evolutionary accounts. Almost all explanations of *why* humans evolved have exalted traditional male tasks such as hunting and toolmaking, and thus mirrored the social milieu in which women were thought to be less important than men in the progress of the world.

We cannot ignore the fact that underlying social creeds are parlayed into beliefs about human origin. Without relevant prehistoric evidence, or without its systematic exploration, there is no choice but to relate the past in a way that ensures the teller's benefit and dominion in the present, real or imaginary. As people of different gender, race, and ethnicity have struggled to find identity in the modern world, elements of bias in the old formulas of human origin have been pointed out and denigrated. The use of origin stories to rationalize sorely desired advantage is a very obvious problem; we have no choice but to be acutely aware of it.

Is there a way to break out of the domineering influences of the present? Can we do anything in our exploration of human origin except mirror superiority or offer rationales for our social beliefs? If we can do no better, any account we offer will be a passing tale, something that decays in time as Ozymandias proclaims his dominion, as things prevailing live out their lifetimes. It will lack something *eternal,* a comprehension of cause and foundation embodied in the true path of our origin.

The tales of dominion prevailing to this day are supported by a powerful sense of separation and difference. Distinction of one thing from another is crucial in matters of domination. I have merely touched on this with respect to race and gender. Inequality, separation, and superiority form a tenacious complex in human thinking.

In the ecological realm, no separation is considered more important than that between human and nature. This girder of human thought sustains the ways we interact with the world. It occurs in the opposites that permeate everyday language. We see the world as either human or animal, artifact or natural. We interpret life in a series of discordant pairs: Culture versus biology. Nurture versus nature. Learning versus instinct. The terms pile up on either side of the wall that separates humanity from the natural world.

Herein lies the seed that interprets history and imagines human origin as an inevitable rise, an ordained proclamation to govern nature—to populate, exploit, or guard it for human benefit. Separation from nature allows human beings to embrace this role, and while we may acknowledge the periodic blows dealt by natural forces, the persistent fabric of human thought claims the dominance of mankind—by sacred appointment, technological achievement, or great intelligence.

The image of human separation from nature denies the genealogy and change that connect all living forms, including ourselves, with other species. If nature's beings are immutable, we possess no connection of kinship with other species, no pervasive link except for the creation of separable beings, with humans accepting the sacred right to be on top. This view, too, has ancient roots in Western thinking.

Is there a way to break this constant round, this tendency to project separation and superiority into the story of origin? The challenge before us is to find an account that holds up, that demonstrates and convinces, across creeds and points of view. Ideally, we would draw a map of the long period of human origin, a map that could be checked and affirmed or altered as it is used. The attempt to discover the events and causes of prehistory means, however, that we are on ever-shifting terrain, constantly changing as new evidence comes to light.

Like the maps of the European explorers of the late 1400s, which mirrored the medieval worldview, our maps of the ancient past still reflect the inevitable pull of our creeds of dominion. With mankind as the center of Creation, cartographers of the fifteenth century portrayed Earth as mostly land. The sea was insignificant in any conceivable plan that placed human beings above the rest of nature. According to prevailing doctrine, no habitable lands could exist beyond Europe, Asia, and the part of Africa known near the Mediterranean; the possibility of land below the equator was denied by the prevailing thinkers. According to Daniel Boorstin, the creed linking man's sacred dominion to world geography comprised the fuel for Columbus's last three trips. He sailed to confirm his map of belief. As he followed it into the Southern Hemisphere, he refused to believe that South America was a major landmass. And he died believing that he had found a passage to the East Indies, not a New World.

Later explorers moved beyond these preconceptions. What they saw, they recorded. Continental positions were charted. The mapmakers refined what they drew. Ultimately, the world came to consist of three-fourths water to one-fourth dry land. And that is how it is, our maps slowly overcoming beliefs of self-centeredness. The spiritual fuel crucial to many people's lives has remained intact.

Our primary source of information about the very ancient past is buried. We can drill into the ocean bottom where microscopic animal shells preserve the moods of Earth's past climates. We can unearth prehistoric sites where different species of humanity thrived on landscapes unlike any in the present. By tedious work we can gather from a number of sources the data of ages long gone. Clues in the ground, external to ourselves, are the only way to challenge our tendency to see human origin as a fulfillment of what we currently are, or what we tenaciously believe ourselves to be.

The search for these clues means a trip into time, a visit to that prologue to the present that we call the past. And we must build techniques of exploration that allow us to observe. The mapmakers did it; so must we.

Humans dominate Earth ecologically. While this is a recent development, it is built upon endowments of the past. The apparent dominion of our species in nature has both philosophical and religious explanations, and a basis in our evolutionary heritage. These grounds, however, usually conflict.

Can we find within ourselves some patch where this conflict, both the evolutionary origin and expressions of the human spirit, can be reconciled? Must humanity hold fast to its presumed separation from and domination of nature? Our understanding about how our species has become endowed with these most awesome and domineering capabilities will define the scope and possibilities in our future. Our view of the world must include an exploration of its evolutionary past. For time and the conditions of Earth itself have fashioned in the descendants of an ape the power to reshape the planet.

CHAPTER II

DOMINION

I

THE desolate moon passes through space in silent view of Earth. Besides rare meetings with asteroids, the latest signs of lunar activity are two-footed tracks that still dint the forbidding, airless terrain. In scattered valleys, heavily booted trails radiate from places where people have alighted. The astronauts observed, collected rocks, and rejoined us on Earth.

The lunar prints made by the weight of men are a curious counterpoint. They imitate a series of bipedal trails three and a half million years older, pressed into a windswept volcanic surface in eastern Africa. These trails attest to the fact that the ancestors of people first walked in the tropical latitudes of Africa. All evidence—the strong affinity of genes between African apes and *Homo sapiens,* the distribution of hominid fossils—points to Africa as the continent holding the deepest record of human ancestry. Now we leave our print on virtually every dry surface of the globe—and in a few abandoned lunar valleys. Between is what we must certainly call "the immense journey," a voyage over land and time. The prints on the moon stand as a reminder of the extraordinary journey of our kind on Earth.

To understand human origin as an ecological process, we must first turn to the current status of *Homo sapiens* as an ecological being. One obvious feature is the presence of people all over the globe. Another is the way we pack ourselves into dense clusters in cities and villages.

Touring many parts of the world by road, one sees mainly people, everywhere. This struck me especially on the island of Java when, a couple of years ago, I first explored its famed fossil sites. By car, my companions and I navigated winding tarmac roads along forested slopes and through the interminable flatness of wet rice fields. All along the

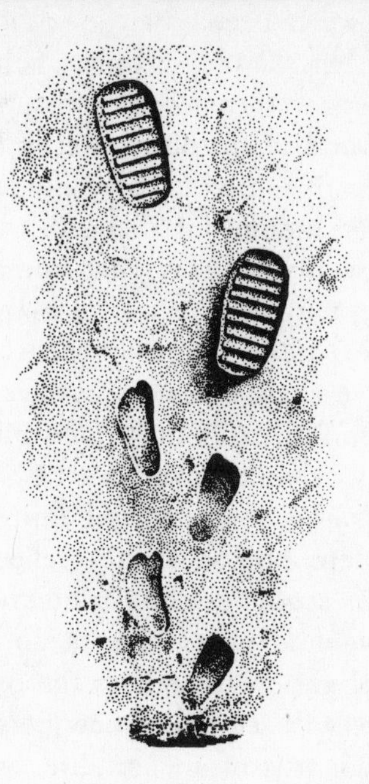

way we saw fellow human beings—in towns and cities and fields of cultivated rice. The population of Java is about the same as that of the entire United States, jammed into an area equal to Great Britain's. People are what you notice from the ground.

Flying in an airplane, you see the human presence in a different form. Over the heartland of North America a mosaic of planted plains and fenced pastures appears, all right angles and radii, copses and clearings. You never actually see the people; just geometry. We have grafted ourselves onto the landscape. Expanding and coalescing with time, the tiles of human ownership have become a vast carpet over the central part of the United States, a thick patchwork quilt of special influence.

These properties of human density and geometry suggest new, vital relations among organisms controlled by people. People apply hand and machine to the land and remodel it to produce food. The result is a new organization, a novel arrangement of plants and animals into fields. It appears to replicate itself across the landscape, and so the geometry of food production is somehow central to humanity's ecological dominance.

We all know that people first grasped the possibilities of marking off and cultivating the land in prehistoric times. By removing the dust from a few thin strata, we can see how the edifice of human dominance has been pieced together. Our examples will illustrate the efforts through time to domesticate other organisms, which led eventually to the unremitting takeover and simplification of habitats.

Whether the exploits of humanity are anomalous, against the grain of nature, is the question hidden just beneath the surface. Our eventual overwhelming impact seems to prove that humanity's ecological status has emerged apart from nature. It is out of place, too awesome to view as continuous with the natural order or with the limited possibilities of other species. Massive change is not, however, unknown in the history of the world before human beings. Dominion must be considered from a perspective longer than that provided by the thin strata in which the oldest burned seeds and domesticated animal bones lie buried.

The ecology of modern human beings reflects an odd turn in our planet's history. As we study matters fundamental to life on Earth, including ourselves, we find that there is no end to the breadth of comparisons and no obligation to confine the proper study of mankind to man. Ultimately, we will see that a new perspective on *Homo sapiens* is reflected in the dry depths of ancient life, even in the once sodden seams of ferns now turned to coal.

2

If human beings were a modest species, we would realize that the greatest revolution in comprehending nature has nothing to do with our own place in it. There would be several candidates for that top attention grabber. Among the most serious would be the tremendous distances the planet's major landmasses have traversed. The continents have moved from place to place over the surface of the globe.

Although sluggish and ongoing, these profound movements are sufficiently dramatic to make knowing them worthwhile. The strange words Gondwanaland and Pangaea are the names investigators have given the supercontinents of the Age of Dinosaurs, when the Americas, Antarctica, Africa, Europe, Asia, and Australia were all attached to one another. That the continents drift and that Earth has reshaped itself remarkably over the eons are facts essentially uncontested among scientists. The geologist's notebook voluminously records the continuity of mountain chains and other land structures across continents now widely separated. Identical or closely related species appear as fossils on distant continents, consistent with ancient connections between these landmasses. And the geologist now knows the currents of molten lava beneath the broken plates of Earth's crust, and how these currents inexorably move the continents this way or that.

Our first lesson in ecological history begins back so far—about 300 million years ago—that the great supercontinent of Pangaea was still being formed. Another 200 million years would pass before Pangaea would first coagulate and then break up into fragments resembling the present continents. It was so long ago that the dinosaurs had not yet offered a hint of their future domination; they had simply not yet evolved. Yet this distant time is strikingly familiar to all of us in a concrete way. The coal you have perhaps held in your hand, that smudges the faces of workers in West Virginia and Pennsylvania, a fuel that has driven our machines and industries over the past two hundred years, had its origin in the swamps of that ancient era.

The landscapes then were populated by stands of lycopsid trees, many of which looked like telephone poles until, late in life, a dense crown of foliage and cones branched out of their top ends, a final phase of growth that expressed their readiness to reproduce. Ferns, in the form of trees and sprawling shrubs, were also abundant, as were less familiar arbors and vines. The death and accumulation of these diverse plants in extensive wetlands of the period, known to science as the Carboniferous, led to the formation of peat. Even today, peat can be cut and dried for use as fuel. And after the enormous pressures and temperatures of 300 million years of burial, the peaty deposits of the Carboniferous took on astonishing energy-releasing properties, able to drive both furnaces and human lives in an expanding industrial society.

We know a lot about the vegetation of this era; fossil plants are abundant in coal-bearing layers. Bill DiMichele, Tom Phillips, and other paleontologists interested in this long-lost history have found out that stable, recurrent communities of plants, persisting as long as 2 to 3 million years, were periodically replaced by new kinds of vegetation. The most dramatic shift took place between the periods known as the Westphalian and Stephanian. At this transition, the abundant lycopsids largely gave way to the ferns. Lycopsid trees had been dominant elements in the wetlands for some time, usually directing their energy into the growth of bark and water-conducting tissues rather than into fast reproduction. This is an aspect of maturation unique to stable environments. Quick reproduction, on the other hand, generally mirrors a response to disturbed habitats where an ability to disperse rapidly and to colonize are important. This latter strategy was displayed by the ferns.

Toward the end of the Westphalian, when a dry phase hit the peat swamps, the lycopsids became extinct. The tree ferns, tolerant of all but flooded habitats, rose to dominance. The weedlike ferns have been with us ever since. Still later, the tree ferns were almost entirely replaced by arbors even more tolerant and faster reproducing, derived from seed plants just beginning to diversify.

The lesson of this period in Earth's history is that ecosystems can hold fast for a long time to a particular appearance—a certain set of species and a specific way of organizing energy—and then change abruptly. *Abrupt* is, of course, within the million-year perspective of geologic time. Precision in the fossil record so long ago is not yet adequate to know whether these painful shifts from stability to disturbance took less than a hundred years or more than a hundred thousand. We tend to think of ancient eras as slow-moving, but that is mere assumption.

Paleontologists studying these ecosystems see evidence of perseverance and stability in the face of climatic change—until something clicked. A threshold was reached, alterations in the biota cascaded, and a new stability came only when different characters, previously minor groups of organisms, began to dominate the stage.

It is significant to note that the Carboniferous drama took place in a foreign theater. Plants and animals assembled, ate, and lived with one another in ways alien to the present. Critical alterations in the next frame of geologic time meant that the modern era was just beginning to take shape.

The Permian is the period from about 290 to 245 million years ago. Before then, the major groups of animals on land were, to all intents and purposes, sanitary engineers. In the Carboniferous, the terrestrial animals were mainly insects and other arthropods who ate garbage. Most of them, like millipeds, fed exclusively on the decayed detritus of plants. Some insects ate live plants, as chew marks in fossil leaves and bored holes in fossil seeds attest; but large four-legged plant eaters capable of massive consumption and heavy damage to living leaves, seeds, and wood had not yet arisen. Prior to the Permian, certain amphibians and reptiles evolved to an appreciable size, several inches to ten feet in length. But these were almost exclusively predators, focused on eating the detritus feeders and one another.

Over the vast scale of the Paleozoic Era, ecosystems on land were unlike those typically portrayed in modern ecology textbooks. Not only were the *kinds* of animals and plants different, so were the channels by which energy was transferred from one living being to another. The sun's energy absorbed by plants was mainly theirs to possess until death. Animals, especially arthropods, entered the food web at that point, consuming what must have been an extraordinary mass of decaying tissue lying on the ground. Energy was passed up the food chain to large animals by one arthropod eating another, by vertebrates eating the detritus lovers, and by vertebrates eating other vertebrates.

During the Permian, one of the most important episodes in the development of life on the continents began. The four-legged herbivores arose. Among the first were the lizardlike caseids with serrated slicing teeth and voluminous body cavities, a combination that equipped them for ingesting large quantities of plant matter.

By about 255 million years ago, four-legged herbivores, most of them reptiles of one sort or another, had radiated across the landscape. At least eight major groups of reptiles had diversified, accommodating their gastronomic reaches to the complex canopies of vegetation that had been in place by the beginning of the Permian Period. The most common plant eaters, the therapsids, moved on rugged legs and snipped vegetative selections from no more than a few feet off the ground. By some reports, this diverse unit of herbivores outnumbered their vertebrate predators by 10 to 1.

Not a single four-legged animal before the Permian possessed the dental equipment to devote itself to vegetarianism. The events of the period thus take on enormous proportions, perhaps most striking in their simplicity. Send out a phalanx of ravenous plant eaters. When all they could stomach were insects and meat, the vertebrates had little if any impact on the success of plants. But with a simple shift in their consuming interests—to leaves and roots and reproductive organs—the Permian reptiles ultimately cast an entirely new light on the relationship between these two kingdoms of terrestrial organisms.

The lives of large animals became entangled with the lives of plants. The herbivores destroyed plants on a selective basis. The plants experimented with ways to defend themselves, with spines, fibrous coatings and toxins, and over time developed ways of making the herbivores disperse their seeds. This new style of interaction had huge consequences for the evolution of plants and plant eaters, while the herbivore radiation created new possibilities in the evolution of predators.

After their explosive radiation, four-legged herbivores became the essential link between the huge diversity of plants and the animals that ate animals. Only at this point did food webs typical of the modern day appear. The entire network of energy, linking sun to plant to animal, was irrevocably altered. Earth has thrived on the tremendous possibilities of this kind of ecosystem ever since, sending out new species and entire radiations of organisms in the eternal reconciliation between passive and active, photosynthesizer and consumer, plant prey and animal predator.

There was substantial extinction on land at the end of the Permian, though it was far less catastrophic than in the sea. Perhaps a fifth of all land plant species were killed off, and as high as half of all four-legged vertebrates. This marked the end of the Paleozoic, the era of old life on Earth. But the juggling of Earth's biota did not budge the new structure of ecosystems established a few million years earlier. The herbivore link in the food chain was here to stay.

The enormous herbivores and predators of the dinosaur clan radiated in the succeeding Mesozoic Era. Their evolutionary booms, followed by crashes and reemergence, were accompanied by a blossoming of flowering plants and subtle developments of small, hyperactive creatures, the ancestors of modern mammals. But none of these critical innovations, not even the terrible extinction of the dinosaurs—the end of the era of middle life on Earth—changed the Permian-style food pyramid to any appreciable extent.

Early in the subsequent Cenozoic Era, forest ecosystems were reconstituted. New climates, new species, and new chunks of continents emerged from the breakup of Pangaea. The herbivores diversified again, and this time the mammals took their turn, consuming the fruits and seeds of an expanding array of plants. At least seventy genera of small mammals arose within 2 million years of the dinosaur calamity. Ten million years after that event, the earliest modernlike primates had evolved—large-eyed, dexterous, kin to the lemurs and lorises. During this same epoch, beginning 55 million years ago, two major groups of ungulates began their evolutionary saga: species having an odd number of toes, represented today by horses and rhinoceroses, and even-toed forms, represented by deer, pigs, and antelopes. The herbivores were present in full force, evolving strange pacts of survival and reliance with the plants and creating their never-ending dance of flight and pursuit with the predators. Change bred change in the continuous reconciliation between reproduction and personal survival.

But the perspective we must reach comes by traveling even closer to the present, near the surface of mankind's origin. Radiations and extinctions ensue. Earth continues its vacillating course. The planet dries out and cools down as the ages roll by. Ice descends from the poles and retreats again. Organisms evolve, new forms emerge—all in repeated reconciliation of the fact that, in time, the conditions of life will be changed.

3

Circuits of energy flow predominant in recent epochs had emerged 250 million years ago. The resilient ecological nexus of the Permian gains our attention for two reasons. First, it shows that species capable of permanently and dramatically reshaping the planet's food webs have previously evolved. Second, the Permian legacy is now being dissected and reconfigured by a new force, the urgent activities of people.

Humankind's need to nurture the landscape and make it theirs resonates a single theme: Cultivation of living things has provoked the growth of human dominion. The current episode in Earth's history was ignited when human populations found ways to produce their own food. The endurance of Permian ecosystems lends proper perspective to the seeds planted by people a mere 10,000 years ago, and to the extraordinary environmental events that have unfolded in our time.

In verdant meadows of northern Khuzistan, people stalked enormous herds of herbivores—gazelle, onager, wild ox—and collected native legumes, herbs, and wild alfalfa. Before a millennium had passed, by about 7000 B.C., the peoples of the area had domesticated these plants, which became ripe after cool winter rains, and the animals that could eat the tough, thorny perennials. According to the archeologists who unearthed the evidence, the people of Khuzistan thus intensified their access to both energy sources, either directly or through an animal converter. Barley, emmer wheat, goats, and sheep were partners in the process.

Around 6000 B.C., people began to build houses with stone foundations, perhaps to underline their sense of permanent possession of the land. Meadows were cleared, and a system of fallowing, which altered the landscape to a significant degree, was developed. Weeds foreign to the region, such as wild oats and rye grass, brought in with imperfectly cleaned clumps of barley and wheat, infiltrated the fields. The weeds thrived on the process of human disturbance, and became food crops later on. After another five hundred years, the people of Khuzistan began to irrigate their fields from nearby streams, and the control of water allowed them to expand to the edges of the river valleys.

As people simultaneously spread and concentrated their mark on the land, the wild legumes and large, free-ranging herbivores were excluded. Hunting and collecting became minor activities. The old uses of aquatic resources were dropped. The plants and animals harnessed by humans comprised nearly the entire diet. Later, cities arose, as did law, writing, complex government, and sewage systems.

And then the people left. As noted by the excavators, northern Khuzistan today is "a pale shadow of its former self—overgrazed, overcultivated, and salinized for many thousands of years." The point of comparison here is not the heyday of city-states, but the bounty of the region at the time when agriculture began.

Similar episodes within the past 10,000 years have been exposed around the globe. In Mexico and along the river valleys of the south-central United States, the weedy triumvirate of squash, beans, and maize was taken under human care. Ultimately, it gave way to intensified forms of agriculture reliant solely on maize. The transition from multiple crops to monoculture took place as complex political systems emerged within Native American societies.

Across Southeast Asia, particularly in Indonesia, two strategies of traditional cultivation developed. In one, the flow of energy toward the human population increased as farmers planted multiple crops, shifted their fields from time to time, and replicated the tropical forest in miniature. In the other, the farmer boldly reworked the natural landscape by packing a single species, rice, into fields of his own making, filled with mineral-rich water flowing from volcanic highlands.

In Indonesia, the colonial government of the Netherlands became woven into the fabric of indigenous ecosystems. Although its mercantile desires spurred the spread and elaboration of rice farming, wet-rice ecosystems were already well ensconced in Indonesia when the Europeans arrived, and had an "inherent growth potential" of considerable scale, according to anthropologist Clifford Geertz. Geertz's comparison with Japan, where no mercantile invasion occurred, provides a clear example of the expansion, without colonial influence, of ecosystems devoted almost exclusively to rice.

Seemingly the potential for human-dominated ecosystems has existed in every place where human populations have been on productive land. In some cases, the crescendo of dominion has been from scratch, at its own pace and manner. Borrowing techniques from neighboring regions has been prevalent in other cases. In still others, colonial contact sparked a spectacular transformation of the landscape, often with shocking effects on the original occupants or an enslaved people.

The human motivations behind the rise of agricultural ecosystems are not easily explained. The circumstances, historically around the world and in prehistory, have varied enormously. Two brief examples illuminate the diverse pathways and effects by which the planet's ecosystems have been altered by people.

The first is the most fascinating and most troubling case. It involves the modification of land in colonial America, specifically in New England. Offering only a brief outline here, I refer you to a startling synthesis of diaries, letters, and data by Carolyn Merchant, a professor at Berkeley, in her book *Ecological Revolutions.*

Beginning around 1642, the removal of wild animals by colonial settlers can be described only as a full campaign. The large predators were driven to extinction. By mid-century, the author Edward Johnson boasted that girls and boys now filled the streets where wolves and bears once nursed their young. Johnson rejoiced in the transformation of "remote, rocky barren, bushy, wild-woody wilderness" into "a second

England for fertileness in so short a space that it is indeed the wonder of the world."

The crops introduced by English settlers were the basic grains—rye, barley, oats, and wheat—and varied root crops and vegetables. Horses, goats, sheep, cows, oxen, and pigs also arrived on colonists' ships. Consuming the nuts of the forests and the native grasses of wild rye, broom straw, and carex, the domestic animals began replacing the dominant herbivores—white-tailed dear, buffalo, elk, caribou, and moose—a process that was largely completed by the year 1900.

Much of what colonial farming introduced to New England undermined the subsistence of Native Americans. Intensive, settled agriculture practiced by colonists encroached upon lands required for Indian shifting farming, which demanded wider acreage. The English cleared forests and drained marshes, which were Indian sources of animals for clothing and food. During the 1770s, many thousands of acres were converted by planting, mowing, and pasturing. These were nodes in a newborn network of commerce. This land was never again available to the native peoples, even though it was integral to their long-term shifting style of cultivation. To the immigrants, it was merely unused and unoccupied land.

Logging of forests provided the backbone of usable energy and building materials for the colonists. Unlike farmers, however, loggers removed trees and returned nothing to the native habitats. A baseline of 95 percent of each New England state was forested at the outset of European settlement. By the nineteenth-century pinnacle of logging, forest cover was reduced to 30 to 45 percent in most of the states. Oaks and hickories were replaced by cattle and sheep grazing on English grasses. Winters were warmer, and summers less humid, by virtue of fewer trees and swamps.

With the arrival of immigrants, pathogens spread. The replacement of wolves and bears by microscopic predators brought massive outbreaks of diseases like measles and typhoid. In 1616 and 1637 epidemics of bubonic plague and smallpox almost annihilated the Indians near the Massachusetts Bay and Plymouth colonies. As colonial population grew, the native population declined, together with its reproductive capacity and ability to retain territory. Meanwhile the colonists' settled form of agriculture required an increase in the labor force and more land to pass on to their offspring. Population swelled. Appropriation of land multiplied. The early 1600s were also the plague years in Europe; in the year 1625, almost one quarter of the population of England died.

In the New World, the cross between deadly microbes and massive loss of native lands brought far greater devastation to the native populace.

The environmental ethic the Pilgrims and Puritans adopted was the "symbolic wilderness." Native systems of land fallowing and multiple-crop planting were joined with long-term European methods. At first, this produced a form of subsistence farming that sustained woodlands. But in short order, land was given out free of charge to America's immigrants on condition that it be "improved"—i.e., planted and built on. The ecological transfer to the colonists was sparked by the belief that the native land was wilderness.

John Winthrop, the noted religious leader and governor of Massachusetts Bay Colony, quoted Genesis 1:28, "encrease and multiply and replenish the earth and subdue it," to argue that Puritans could lawfully expel Indians because they "enclose no land, neither have they any settled habitation nor any tame cattle to improve the land by."

By the 1700s, the symbol of Eden was extended into the New England countryside, since the biblical garden was the first given to humankind to possess and to till. In the transformation of the land by colonial farmers, nature itself was brought into the realm of pious belief. In Merchant's words: "Stones and trees from the realm of wild nature became instruments of culture when converted to fences that ordered the land. Culture itself was the nurturing of food and fiber." There was a "struggle to set humans apart from nature."

In a few generations, this first transformation was complete. Merchant recounts: "In 1735, twenty-seven surviving Pequot Indian men complained to Governor Joseph Talcott of Connecticut. They were now dependent on cattle and corn since so little hunting land was left. The previous spring their corn had been destroyed by English cattle and their apples eaten by English swine. . . . Captain James Avery, a justice of the peace in New London County who investigated their complaint, concluded, 'I have enquired of said indians, and can't find that there has been any real damage done them.' "

By the nineteenth century, the doctrine of agricultural improvement meant social progress by cultivating the land. It required good accounting and farming for money. It demanded awareness of the latest scientific reports on soils, crop yields, and profits. A market system of food distribution was born, fueled by industry and new forms of energy and mechanization. The burning properties of coal were discovered. Iron was produced and employed in new ways. The fossil-fuel age had begun. And with the industrial redirection of energy, traditional farming gave way to intensive agriculture and specialized production.

As markets became intimately involved with farming, people's attitudes were transformed. Scientific management of farming meant, in the eyes of Americans, a laudable increase in human control over nature. Saw, textile, grain, and iron mills required power, which was created by damming streams. The belief that stream flow was a part of nature was now irrelevant, and the older concept of water rights, whereby people must not modify the flow of water to others, was perpetually altered.

Motivated by business sense, the lumber industry began to replant forests, and the forests began to recover—as objects owned by people. Lakes and swamps were not allowed simply to exist; even the ethic of conservation was centered on serving pragmatic, human needs.

Merchant's treatise dissects the complicated factors underlying this second ecological transformation, an epidemic of industry and commercial croplands. Size of farms, land inheritance to sons, decline of soils, use of fertilizers, and the intricate connection between offspring and farm productivity all were part of the fabric of cause and effect. Merchant's research also documents a growing psychological squeeze on farmers. As market-oriented farms expanded, people entered a time-limited vortex of mental energy and things to do. Families began to feel more isolated from their communities. Farm work and economic management of the family became the central goals of life. People were now too busy. They were growing apart while simultaneously becoming staunchly connected in matters of commerce and information. The stage was set for the emergence of modern American society.

From one perspective, human dominion arises from a single track, a course fueled by Western notions of economic growth. Adoption of Western technologies and approaches to economic gain is the popular reason given for the spread of human-dominated ecosystems. This viewpoint is now widely held and largely unquestioned.

Yet, as suggested by Khuzistan, Mexico, the river valley chiefdoms of the Iroquois, and the rice systems of Asia, the conversion of landscape to strictly human purposes has diverse histories, reflecting the multiple cultures and places in which this endeavor has been tried. It is tempting to focus attention on the economic ethics and industrial technologies now capable of altering Earth profoundly and rapidly. But by surveying the proliferation of agriculture and of human imperatives worldwide, we see that ecosystems never pressed by global markets, never heated by fossil fuels, never affected by colonial imposition, have been transformed according to a general pattern. Even today, economic notions and techniques traverse lively cultural membranes and attach to receptive centers in the diverse approaches brought to life by human beings.

Before I attempt to summarize that pattern of change, there is a second example, drawn from my experiences and conversations with farmers in Kenya. It shows how the ideas of human dominion catch fire—and the collectivity of personal decisions that propel them.

The men I dig with grew up as farmers. All of them still are; each one owns a rural *shamba,* or cultivated plot. For the most part, these men hail from Machakos district, east of Nairobi, and travel each year to work in Maasai country at Olorgesailie, southwest of Nairobi. Their first language is Kikamba, although they almost all know Swahili and English as well. I remember one hot, glaring morning, which started in an excavation pit with the foreman of our crew, Muteti Nume.

During a break, Muteti began to tell me about digging his *shamba.* By loosening rocky soil, he has worked hard to grow maize and beans, and has collected awards for the improvement of his *shamba.* "Improvement" was the word he used. He spoke of planting tomatoes, onions, and other cash crops. "I like to try new things," he said. "It is my family who benefits."

I was struck as he talked about "my soil," "my rocks," and "my trees." Apparently, ownership is assumed wherever people have adopted the planting of crops, regardless of culture or external ethic.

"Before my grandfather's time," he said, "many kinds of animals lived in the area. Lions, hyenas, giraffe, buffalo, zebra, many gazelle. The Kamba people then were mostly tending cattle, and they moved from place to place. After a while, my ancestors found some good areas, settled down, and started to build *shambas.* Whenever the people came to a spot they liked, the men would go around and chase all the animals away. All kinds, even the antelopes. Because any kind of animal can disturb the crops and the people who live there." He spoke of a time before colonial influences swept into his area.

The sun was already high, and Muteti broke off this conversation to inspect our excavations to the west. No sooner had he left than Francis Musila, another member of our team, arrived. One of Musila's responsibilities is to drive to the nearest village, Oltepesi, and fill our large water drums. Francis is happy in the bush when people are around to discuss the issues of life; he thrives on observing the world and talking to others about it. He confirmed what Muteti had begun to tell me.

"The people of Ukambani," he said, "used to rely on meat, animal blood, and milk, like the Maasai have always done."

A century ago, many wild animals thrived in his area: "People were not as many as today, and animals could hide in the bush and not be disturbed."

He recounted the removal of wild animals by the farmer. Because birds eat grains, they are dispatched with a slingshot. Monkeys are killed with bow and arrows. "These animals are clever at stealing the crops. Keeping them away can occupy a farmer and his dog all day. All the insects and antelopes are predators. They eat crops before the farmer can harvest. The mouse is clever because he waits until the farmer takes the maize into his own home. Weeds destroy the crops. You can never dig out everything; there will always be weeds in the ground."

I asked Musila about irrigation and water rights. He explained, "Those of us in the hills take water from the river using hosepipes. But this lowers the water, and people argue about that. Someone will say, 'It is my turn.' And someone else, 'Lend us some water, please, because the people above have taken it all; those of us below have none; there is none in the river.' So people agree that those high up take half, leaving half for the farmers below. Yet this is hard during a drought. People will always be thirsty for water. Even though water flows downhill, the water won't pass into the valleys below, because of people."

Musila was way ahead of my own ideas about how the farmer and the society in which he lives lay down the seeds of agricultural ecosystems. He sketched for me the maze in which the Kenyan farmer lives. People now have more babies, which means more sons among whom to divide the traditional inheritance of land. A mother and father, he calculated, now have about six kids, so the average family size is about eight.

"But some families are larger than that. In my family we are three brothers and my mother. My mother is alive, my grandmother is alive, my stepmother is alive with five boys. Okay," he said with a nervous laugh, "I have seven children, but one is married already; so there are six. Six, my wife, and the others all come under me since I am the eldest child in my generation." He stopped to think. "So five plus six plus six; we are altogether seventeen at home, and my wife and myself.

"The land acquired by our ancestors is getting finished. Anything less than ten acres would be a very little piece of land. Some years back, a person didn't bother owning such a large area. Our fathers kept more cattle, but now only ranches possess a big area of pasture. The rest of the good land is for our *shambas.*

"Now almost every young man must buy his own land somewhere else, sometimes far away from family. Many farmers think about jobs away from the *shamba,* which can bring in money for land. The mother is looking after what the children will eat today, and the father is looking after what the children will eat tomorrow. She goes to the market, takes care of much on the *shamba,* and comes back to the *jiko* [kitchen fire]. The father will go out to work and try to get money. Our children face the same matters. But education needed for a job requires school fees. To get money, people will sell pieces of land. The pain comes because that land is the inheritance of our children."

Musila explained to me how food, land, education, and money reflect a series of choices, variables in a complex equation. "One can pay off or buy the other," he said. The Kenyan farmer chooses between growing food sufficient for his family, or growing an excess for the marketplace. Some people have now converted all their living into crops for selling. They manage bigger and bigger pieces of land. "Many people are thinking about businesses besides the *shamba.* And so there are choices between working hard on the *shamba* or away in the city."

Commitments and life choices tighten on the Kenyan family. Certain pathways become pitfalls; sometimes there seems to be no place to turn.

At one time or another I've asked all of the older men on our crew, including three who are more than sixty, why they spend several months each year working at the excavation, far away from their families. Each one answers: We must have food; the extra work guarantees it and gives a chance to pay for our children's school.

I asked Musila if life wasn't better when the only concern was to grow enough food for the family. Wasn't it better without the problems of the marketplace, making a business, or working away from home?

He mused a second and replied, "Now days are better because we had nothing to defeat famine before. Famines used to come about every ten years, when the rains failed. People ate roots and soil. They tried anything as food. No marketplaces. Only villages and farming for yourself. There was nowhere to turn when people were really starving. People just waited for death, waited for the hyenas and dogs, too weak to chase them away. This was the 1800s. Famines then were worse than today. Now we have advanced ways of farming. Moving around is very important in times of famine. If there is wheat in Nairobi, brought by the airplane, someone can go get it. If there are mangoes in Mombasa, someone can go for them. Farming is a business because of hunger, trying to keep it away."

To be sure, Western influences tug on Musila. Outlooks and examples from the outside affect whether he tries to expand his *shamba* or works outside of it, with me. Yet Musila's answer is not a simple projection of Western values. It mirrors the distinct realities of life long experienced in eastern Africa and many other places around the world. Staving off famine is deep in Musila's history, and in every Kamba farmer, as they search for land, give something to their offspring, expand the fields a little farther. This is life, personal and deeply felt, and the result is collective.

4

In the preceding portraits, we begin to discern the distinctive paths in the expansion of an agrarian life. If we were to face the entire array of cases—in prehistory, written history, and the present—we would reach a simple deduction: The collectivity of *Homo sapiens* has ignited the most extreme shift in ecosystems since plant-eating animals first arose on Earth 250 million years ago. We remodel the planet's landscapes, simplifying them as our method of improvement. Ancient food webs lie in tatters as we nurse heartily upon their energy.

Cultivating food is the fundamental sign of human dominion. In a brief 10,000 years, human beings have transformed ecosystems by burning, planting, hoeing, and watering. By laying out the general pattern of alteration, we can see just how substantial the human effect has been, and just how inexorable its spread is.

The planet has faced the traumas of asteroids and droughts, and extinctions massive and minor. Through it all, the flow of energy has stayed remarkably consistent since the Permian. In that epoch, the food pyramid took on the shape featured in any basic text on ecology. Plants were rooted at the enormous base. Herbivores inhabited the narrower middle. The meat eaters occupied the small tip. Carnivores have always been the least abundant, for they are the organisms farthest away from the solar source of all energy, standing in line behind plants and plant-eating animals.

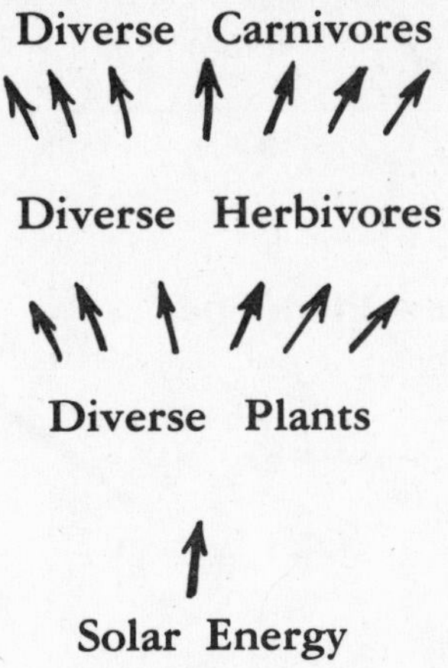

Diverse Carnivores

↖↖↖ ↑ ↗↗↗

Diverse Herbivores

↖↖ ↑ ↗↗↗

Diverse Plants

↑

Solar Energy

In the drawing above, energy flows up stepwise through the food chain. The features of the physical world, such as rainfall and temperature, largely control the nature of vegetation. Plants have a strong influence on the diversity and adaptations of the plant eaters. And the number of herbivores—the potential prey—determines the character of carnivores in the ecological community. The top carnivore is the animal at the apex of the food pyramid—a predator on other animals, a prey to none.

Like the flow of energy itself, control over the whole system appears to move up the food chain. In actuality, a downward effect also figures prominently in the workings of ecosystems: Predators select prey according to certain attributes, casting an unwitting vote on the survival and evolution of prey populations. Even the plants assert an influence over the very features of moisture and heat that guide their own limits of tolerance.

Hardly a person on Earth lives in a food chain of this sort anymore. People have constructed and now govern their own food chains. During a conference in 1987, it dawned on my colleague Robert Foley and me that agricultural ecosystems all over the world have a very different foundation. Their operation is distinctive. In casual conversation, we sketched some of the fundamental differences.

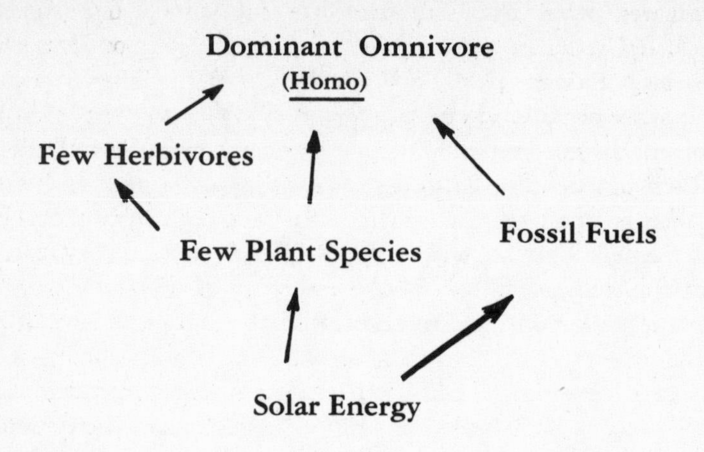

Humanity's grasp on the food chain is organized around the control of plants and animals—their reproduction, yield, and diversity. The greatest energy and strongest controls do not move up or down between vast fronts of animal and plant species; they move in relation to a single species, human beings.

In every agricultural center, a very few of the plants in that region become obvious candidates for domestication. Breeding and growth come under human command. The diverse array of noncultivated species is gradually exiled, even though in many cases these species once helped to replenish soil nutrients. Instead of the cornucopia at the base of the old food chain, the tender care of fields divides and diminishes plant diversity.

Human beings are not equipped to digest grasses, yet we have taught ourselves to breed and process certain species, the cereal grains. We make them edible, rich in protein, high in energy value. All we can do about other grasses is to stand by while various animals consume them. So, in an equally clever act, we put aside the grasses we can't eat. We clear the land to create pastures. Then we allow a few well-bred species of herbivores to eat those grasses for us. And in one form or another—meat, milk, cheese, eggs, and so on—we eat those herbivores. In utter disregard of Permian accomplishments, we eliminate other plant-eating vertebrates, especially those that consume our crops, or compete for the pastures of our domesticated beasts. While still a vital link, herbivores are often reduced to a single species that serves the energy needs of the human population.

Carnivores other than ourselves are completely discharged from duty—killed at a rate proportional to their threat upon domestic herbivores and human beings. With increases in human population density, other organisms have unexpectedly taken on the predatory role. Viruses and diverse vectors of disease weave microscopic ambushes, pursuing their innocently pernicious ways at the cost of human life. People have taken it upon themselves to declare medical warfare on this uninvited element that has taken advantage of human ecosystems.

Over the decades, debates on the origins of food production have soared and stalled. According to one set of theories, a more sedentary life caused the population to increase and created a threshold—a critical density of people on the land—that spurred the cultivation of foods. An opposing set of theories has also been presented: The beginning of food production came first, creating the conditions that made the nomad stop, settle down, and quicken his rate of population growth.

Whichever way we point the cause-and-effect arrow, it is clear that food production and population growth have proceeded together. In different cases, either population size or environmental change may have tipped the balance between collecting wild foods (when resources were abundant) and intensifying the work effort by tending crops (when resources were low relative to population size).

Regardless of the exact sequence, people coalesced. Cities arose. The business of life grew even more dependent on the yield of farming. As cities matured, so did the economies of agriculture. Farmers and society as a whole intensified the production of food, and the new ecosystems, dominated by people, sometimes became committed to a single crop extending beyond sight. Fields composed chiefly of a single species of grass replaced forests, diverse wetlands and prairies. These ecosystems—the monocultures of wheat, corn, and other cereal grains—began to emanate from the centers of human agricultural activity.

Ultimately, large cities and machinery drew on new sources of fuel, which reinforced the energy circuits already centered on people. As if our insult to the Permian were not enough, in a new and defiant act we resurrected the Carboniferous and harvested its coal. Before too long, we would release even older forms of life, drawing heavily on the deep wells of petroleum, energy long held dormant in dried Paleozoic seas. Ancient energies circulate through our homes and run our factories; they lubricate and push machines used in our massive transformation of landscapes, enabling millions of acres to be converted almost entirely to human use.

I call these new habitats *human-dominated ecosystems.* Human beings, a single species, now control food webs of an entirely new shape. We have remodeled the terrain and the underlying pathways of energy exchange. Humans are top-carnivore and top-herbivore. We strive to become the first and last creature to prey upon fields of plants and herds of animals. We ensure that no creature preys upon us, and none does except ourselves and the tiniest of organisms.

As Indonesian swiddens and people's gardens around the world attest, not all farming entails monocultures and massive loss of diversity. Wild plant and animal species coexist in many areas with crops and livestock. But the biomass and range of undomesticated species have shriveled in the presence of human cultivation. Remodeled landscapes comprising a small number of staple crops and animal herds are encroaching most rapidly, and with the greatest influence, over Earth's terrain.

According to a 1988 article by Dr. Mark Plotkin of the World Wildlife Fund, "less than 20 plant species produce most of the world's food," and "the four crop species—wheat, corn, rice, and potatoes—feed more people than the next 26 most important crops combined." In a given year, government policies and Western-based industries may lay claim to thousands of square kilometers of forest for agriculture, mining, or urban expansion. And a thousand farmers may make up their minds to plant crops on a few more acres. The transformation occurs at all levels.

Human-dominated habitats now occupy every climatic zone where there is arable land, and every continent except Antarctica. According to estimates well over a decade old, as much as 15 percent of the total land surface on Earth is devoted to agriculture. Its rapid invasion of new terrain and efficient replacement of other ecosystems assures us that this figure must now be higher. Human-dominated ecosystems have taken over much of the most productive land on the planet.

Huge urban sprawls are already a vital part of these ecosystems, though little food is grown there. They are the black holes that draw massive amounts of energy, the enormous demand that distills the food-production ecosystem to its purest, most rapidly expanding state. The simple principle here is that urban people, all of them, need to consume the products of the land.

5

As human-dominated ecosystems enlarge, their side effects race over the biosphere and permeate the planet's surface. Loss of species, disappearance of habitats, thriftless pursuit of resources, fumes of industry, and the effluence of a global populace are documented exhaustively in books, articles, and newspapers.

Population growth is a bellwether of human-dominated ecosystems. A population boom means that people aggressively adopt the land around them for exclusive purposes of agriculture, fuel and mineral acquisition, and waste disposal.

Population is growing by over 90 million people a year. The current number, about 5.5 billion, represents a doubling of the world figure since mid-century. The Worldwatch Institute's report for 1992 observed that the world's urban population would double in just twenty-two years, while at current rates the cities of developing countries will do so in just fifteen. Croplands will inflate; so will the technologies applied to a given piece of farmland. In the troubled southern Sahel of Africa, human population has more than doubled since 1960; cropland has expanded by a third.

The number of animals under human control is also growing. The population of cattle, pigs, and other domesticated mammals has increased from 2.3 billion to 4 billion since the mid-1900s; domestic fowl have swelled from about 3 billion to 11 billion.

As population grows, so do the industries that support human dominion. Agricultural expansion now has minor impact compared to the growing demands for energy, metals, minerals, and the disposal of industrial spoils. During the past twenty years, world energy trends have oscillated. Oil and coal usage has flipped up and down depending on economic, political, and ecological factors. Oil consumption is up 12 percent since 1973. Output has declined in the United States; certain oil fields in Russia are depleted. It's projected that within a few years, the world's oil needs may require all-out production by the Persian Gulf countries, a significant dip into two thirds of the world's proven oil reserves.

Worldwide, 30 percent more coal is consumed now than during the mid-1970s. Natural-gas use has risen 87 percent in two decades. Fossil-fuel use in the prime industrial nations is not about to level off, and in the Third World it grew from 18 percent to 28 percent of the world total from 1970 to 1991.

The effect of massive fuel use is magnified by the exhausts it creates, which alter the air and water. Greenhouse gases—including carbon dioxide emitted by burning coal and oil—and CFCs (chlorofluorocarbons) from foam containers and refrigerator coolants—increase in atmospheric concentration each year. The former affects global warming; the latter depletes atmospheric ozone.

The search for new energy sources has led to the production of 100,000 metric tons of nuclear waste. Millions of tons of radioactive tailings worldwide are left from uranium mining. None of the proposed burial sites—under deep geologic layers, the ocean, the Gobi Desert—is ready yet. Half of the irradiated matter will be stabilized by decay in about 77,000 years.

Besides fuels, the mining and processing of minerals has contaminated aquifers and rivers for centuries. A study in the early 1990s claimed that 2.7 billion tons of waste, much of it hazardous, results from mining and smelting each year. These activities, so essential to the modern world, have dramatically changed the soils and biota on huge tracts of land. The tendrils of human-dominated ecosystems reach far beyond the planted field.

The spread of our ecosystems also calls for a dramatic rise in the basic production of things, from food containers to freezers, newspapers to shopping malls. Where do we put it all after we're done with it?

The commonest thought on this matter is obvious: As long as my home isn't spoiled, things can't be too bad. I am safe. The question remains, where do we put all the waste, all the people, and all the land to feed the people? The dire writings of Edgar Allan Poe and other conspirators of horror arouse my imagination. But no story evokes such odd images of gloom as that written by the collective corpus of *Homo sapiens.* Widely perceived as progress, the work of advanced peoples, our dominion is full of irony—even if it does not suggest our demise, as we will later observe.

There is, naturally, wide debate about how important these matters of resource use and waste disposal are. Climatic data strongly indicate that Earth is warming, corresponding to the emission of greenhouse pollutants. Based on atmospheric measurements and the reactive laws of chemistry, the ozone shield is thinning in places, allowing more ultraviolet radiation to hit Earth's surface. Our concern, of course, is mainly about our own skins—not whether there might be larger implications for the environments around us.

Some authorities doubt that global warming is occurring and

whether the data require any critical action now. Other hard thinkers believe that, regardless of data, Earth is a maternal source of balance, maintaining nature's life-sustaining equilibrium. Like Mount Pinatubo, a volcano will erupt or some other factor will come into play, cooling Earth to maintain temperature stability. A two-year reversal in the trends arising from human-dominated ecosystems is raised like a victory flag for this thesis. Finally, a different voice, heard far and wide, preaches an immodest creed: Regardless of minor side effects on ozone and temperature, air and water, using the planet and improving the land are symbols of our sacred trust, the dominion over beast and fowl given to us by God.

The axes of debate about the environment are multiple; the distinctions in viewpoint are complicated. It is difficult not to be awed by the asteroidal-size disruptions that a single species has been able to create. There is no escaping the extreme impact people have had compared to any other living being, or to any of the symbols of prehistoric life. The many species of dinosaurs ruled because some of them were large and abundant. The woolly mammoth and Eurasian rhinoceros also had their later heydays on the northern steppes. But if that is the kind of dominion evolved in nature, how did human beings come by theirs?

Our remaking of the landscape seems so different. We remove forests and create immense fields. We embrace the plains and place upon them our own species of grass. We make wetlands dry, and, by irrigation, drylands wet. We create deserts by the grazing and foot pounding of herds under our control. Our desire for crops exposes hillsides, which then enter the sea by erosion. The oceans bear the unmistakable imprint of a terrestrial biped.

Looking at these endeavors, we cannot mistake our impact on species. Pesticides and industrial exhausts have long been recognized for their stranglehold on the biota of lakes and barrier reefs. Land organisms are rapidly lost by the dispersal of human-dominated ecosystems across continents and islands. Based on recent extinctions in rain forests, botanist Peter Raven estimates that one quarter of all tropical plant species will be gone in three decades. Calculating the minimum rate of extinction, biologist E. O. Wilson estimates that at least 27,000 species are lost each year, on the order of one out of every thousand species that exists on Earth.

The insects, super-diverse by any estimation, account for much of the current crash, the result of human diminution of tropical rain forests. But no major group of animals is without peril. Populations of mammals in Australia, fish in North American lakes, amphibians, reptiles, and birds are collapsing worldwide. The wild cats, bears, and the biological group to which people belong—the primates—are among the vertebrates most precariously perched.

In soaring speculation and loving fantasy we hope to commune with life-forms from other planets. But there are intelligences on Earth with whom we share exact copies of almost all of our genes, about whose survival in African forests we are silent. An estimated 550 mountain gorillas remain in the wild. Bonobo chimpanzees are more numerous, an estimated 10,000, all within Zaire. Human populations contract around the habitats of these species. How effective will wildlife parks and reserves be against the forces of human acquisition, civil unrest, and worldwide alteration—the spread of our ecosystems?

In a popular book published a few years ago, Bill McKibben denounced the "end of nature" caused by human beings. He wrote about "the need for pristine places, places substantially *unaltered* by man," and argued that the fundamental symbol of nature valued by humankind is that which is distinct from the human realm. McKibben sadly noted that pristine nature has been overwhelmed by human action: "*[W]e have ended the thing that has, at least in modern times, defined nature for us—its separation from human society.*"

His book repeatedly demarcates something *original*—a brook, a hillside, an entire planet—that was present before human dominion: "We have killed off nature—that world entirely independent of us which was here before we arrived and which encircled and supported our human society." He then describes a "new nature," a human domain of unknown change and distraction: "The salient characteristic of this new nature is its unpredictability, just as the salient feature of the old nature was its utter dependability." Although there are rains and sunshine, "on any larger scale nature has been quite constant, and on a global scale it has been a model of reliability—'as sure as summer follows spring.' "

This concept of nature, original and constant, is widely felt in the literature and lectures on environmental conservation. There are hints of it in the statistics on habitat loss: "Home to at least half the planet's species, tropical forests have been reduced by nearly half their original area. . . ." Concerning temperate rain forests: "Of the 31 million hectares once found on Earth, 56 percent have been logged or cleared."

Forests, wetlands, and other threatened features of the natural landscape are construed to be original elements of Earth—eternal, except that they are now transformed by the efficacious hand and uncontrolled waste of mankind.

The desire to define human impact on plants and animals offers a further example. Biologists seek a way to determine the rate of extinction before people dominated the scene. As a reasonable guess, E. O. Wilson proposes that one out of a million species each year normally became extinct before human interference. And he adds, "Human activity has increased extinction between 1,000 and 10,000 times over this level in the rain forest by reduction in area alone."

We need to know the original condition in order to recognize the biological cost and future perils of human activity.

My point here is not about caring, for that we surely must. Our central question concerns the way in which this extreme situation came about. How is it that humankind arose within a stable natural order, yet creates such profound instability?

Following the central tenet of evolution, species evolve in relation to external conditions. If, in fact, humans emerged within habitats marked by stability—balanced and full of species—how is it possible to make sense of the ecological disintegration of nature that has been uniquely brought about by human beings? We are caught in a strange quandary between nature's constancy and our own actions, which throw it out of balance. In evolutionary terms this means that the process of natural selection somehow equipped us to become drastic modifiers of the very conditions in which we evolved. Confronting this quandary, we find that for over a century, the science of humankind has been stretched between two visions regarding the conditions of our origin.

The first assumes that environments were slow to change before human dominance. Human ancestors evolved in the stable habitat of the forager, in tune with the surroundings. These predecessors lived within the status quo of nature, their emergent humanity fueled by the intrinsic properties of tools, culture, and brains. The environments of tropical hunter-gatherers are believed by many to be a model for the habitats in which we evolved, places where the ancient steady state of natural selection might still be studied. On the other hand, E. O. Wilson proposes: "Human nature is . . . a hodgepodge of special genetic adaptations to an environment largely vanished, the world of the Ice-Age hunter-gatherer." While there is debate as to the exact kind of environment, the scientific venture seems bent on finding the single primordial setting in which hominids were nudged toward humanness. Perhaps not by coincidence, this outlook is crisply mirrored in the concept of *original nature* that arises in response to the current environmental crises.

The second vision, complementing the first, touts the directional manner of natural selection and the runaway effects that resulted in the human lineage. Opening virtually any text on human evolution, we find: "The rate of change due to selection is usually very small each generation, but small changes can have great results when they proceed in the same direction for long periods of time." This description comes as no surprise. It is a straightforward restatement of Darwin's original proposal.

Every account I know goes on to define the direction of change. According to anthropologist Richard Klein's popular text, protohominids became more terrestrial as the spreading savanna inspired greater reliance on both tools and meat. At least one early hominid lineage responded strongly to these selective pressures, and in this lineage, ground living, tool use, and meat eating interacted to produce a dramatic increase in brain size. Again, this statement provides no surprises; adaptation to the savanna has been the dominant scenario in the science of human evolution ever since Darwin.

Gradually, the forces of directional change engulfed what was to become the human lineage. As trees were irreversibly replaced by grassland, the fundamental features of humanity were carved out and joined together.

Directional change and selection pressure comprise the lens through which our current ideas about human evolution are viewed. Stability, integrity, and *original* environment are the bold strokes in our prevailing portrait of nature prior to human dominion. We are caught in a whirlpool of assumptions: Human beings arose in a single-minded state of nature; in that context natural selection laid its hands on our ancestors and gave a continuous, directed push toward the human condition.

But the gradual increase, the increments honed by natural selection, do not add up, do not resolve our dilemma. We remain the dissonance, the note of discord with the conditions of our own origin. The result, by some irony, was a freak of evolution—separated from nature by an act of nature. *Homo sapiens* causes rampant extinction, in conflict with his background state. He alone creates massive change and environmental instability. He is undisputed landlord in ecosystems of his own making. And he holds a defiant upper hand over the planet.

We are stuck. Human domination appears to revolt against not only forests and rivers but also our evolution within natural environments, and thus against the idea of our connection with them. Dominion remains the obvious kernel of human existence on Earth—and it continues to defy explanation.

The root of this conflict lies much deeper than the anthropological sciences. From conservation movements and philosophies of human conduct alike, we hear that "humanity is a product of nature," and that this fact must be borne in mind before we bring about a global ecological catastrophe. Yet the sciences of humanity and evolution have provided no grounds on which to reconcile the product-of-nature assertion with the forewarning. The impending catastrophe lies in our cultural activities, we are told. And so philosophies and sciences both speak to us of culture and nature as opposing forces. What many consider the fundamental human adaptation is thus held apart from nature—in fact, opposite to it. With this divide, the idea of human derivation from nature is as distant as ever.

Moral creeds about human harmony with nature stir within many of the globe's sacred traditions—the Christian, Buddhist, and Native-American, for example—but what is it exactly that causes human beings to exploit and to disrupt? What forces of friction between people and habitats are the creeds of harmony attempting to mollify?

There is more than one approach to the conflict between human dominion and nature. First, we might accept it at face value: Dominion is *not* consistent with nature. The planet's environments were stable until relatively recent alterations by people. Efforts in the biological sciences to determine the parameters of nature and its state prior to human interference are hallmarks of this approach. Therefore dominion arose as a progressive outgrowth, an evolved separation from natural habitats. Culture is the lever that has pried us away from the primal conditions of origin.

Another approach casts serious doubts about human separation from nature, in both the present and the past: The environmental saga of human origin led *Homo sapiens* to engage in its current imperatives of acquiring resources and altering habitats. The capacities for dominion are somehow consistent with the history of nature and human origin within it. This approach is equal to the first in its embrace of the evolutionary process. Its outlook is, however, very different.

In what follows we will dismiss the idea that nature is stable and slow to change, while humans alone are disruptive and in conflict with nature. The first approach has false premises about the course and causes of human origin. Our lineage arose as large, periodic fluctuations governed the conditions of survival. The effects were felt by the hominids and experienced by the biotic world at large. The distinguishing qualities of human culture emerged later, attached to extreme, repetitive shifts in climate and biota. The power to alter our surroundings grew stronger as a way of moderating erratic environmental change. The results were intimate and consistent with nature's own periodic face-lift. This outlook on our evolution differs sharply from the tenets about nature and humanness lodged in Western thought.

In every epoch, the prevailing view of mankind's place in nature has emerged with human action and spurred its impact on Earth. Different outlooks are likely to create different viewpoints and approaches to life. Recognizing our origin as an ecological genesis will affect what we conceive of doing next. It is important that we get the relationship between nature and humankind *right,* both in its true long-term development and in its present possibilities.

By the fleeting addition of minuscule effects, the years will pass. A time will come 250 million years from now—there is nothing we can do to prevent this. And our era will be as distant as the Permian. Humanity will have been but a bridge in a swarm of helixes, having passed the grace of life into unknown intervals of time. Perhaps other contemplative beings will have evolved by then, built upon our ancestry or as distanced as possible from it.

Eventually they will meditate on that dim period of alteration, as distant from them as the Permian is from us, and they may wonder about the influence of a bipedal builder upon their destiny. Their kind will be so profoundly connected with nature as to be born in it, however nature then appears. They will wonder whether we knew similar things.

At present, what we know confidently is this: The world now rests, more than at any time since its birth, on the legacy left by a single species. We are that species; in this lies our unexpected immortality.

CHAPTER III

NATURE'S ALTERATION

I

THE hottest, most desolate place I know is called Lainyamok. It is a Maasai name that means "the place of thieves." During four months in 1984, my colleague Pat Shipman and I excavated there. We saw the mercury halt at the top of the thermometer, and we thirsted for signs of early human activity. Lainyamok is in southern Kenya, some rugged miles west of a lake known as Magadi.

It was at Lainyamok that I had my first true encounter with a kind of dynamism in the earth that might initially appear to exaggerate the vitality ascribed to a passive pile of silt. A careful eye searches for multiple clues whose strange union may reveal spectacular events in dull dirt. The clues at Lainyamok were of a large old lake, bigger and older than Lake Magadi, which had abruptly dried up nearly 400,000 years ago. The pink silts it had deposited were a kind of tombstone, marking the death of the lake. They were laced with deep cracks, developed as the lake dried, first to a mere wet muck, ultimately to an arid wasteland. Intense drought enveloped the area. Calcium carbonate had filled in the cracks, evidence of extreme evaporation. And chemical tests showed that just prior to its death the lake had possessed one of the highest-known levels of salinity and alkalinity of any lake in eastern Africa, past or present.

People weren't responsible for the destruction of this lake. Leaving behind a thighbone, three molar teeth, and some stone tools, the bipeds with large brains had to live through it.

A year later I began to organize a research expedition to Olorgesailie, forty kilometers northeast of "the place of thieves." In light of our curious observations at Lainyamok, I began to ponder seriously the environmental passages of the human lineage: Did the ancestors of *Homo sapiens* experience long stable environments and adapt to them in a progressive manner? Or did they face frequent changes in the conditions of life and survival? When alterations occurred, were they abrupt and extreme or slow and gentle? How easy was it for our hominid forebears to hone a single, stable way of life?

Olorgesailie is a small area draped with fossil deposits. Its chronicle is not particularly old. The first sediments were laid down there a little over a million years ago, so its prehistoric annals do not stretch back nearly as far as the oldest roots of the human lineage. Yet its stratigraphic archive provides a crisp record of the past million years; and its display of lithic implements and fossil animals that bear on the ancient ecology of humankind is continuous and accessible.

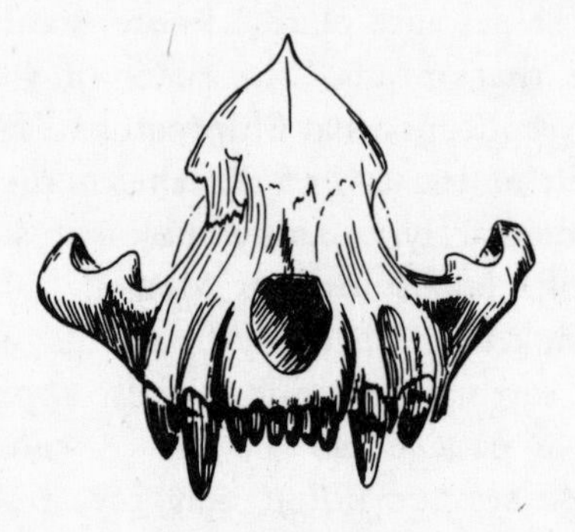

The Olorgesailie region is ringed by a huge brown edifice of volcanic rock known as trachyte. These walls of hardened lava are ominous remnants of dynamics far exceeding even the rich texture of life in this area today. They represent outpourings of volcanic rock, huge disgorgings pressured outward from beneath the ground, evidence of the planet's ever-present growth pangs. Between 1.4 and 1 million years ago, molten rocks blanketed the terrain as the guts of the rift released their hot burden onto the surface. This was early in the time when we know, from chipped stones, that early humans walked the area. And it is here, as in other places in the rift, that our ancestors felt the tremors of a mutable Earth, telling of a place meant not just for human purposes, but, rather, occupied by the vicissitudes of nature. By direct experience, they knew this better than we.

A little over a million years ago, rain and streams poured onto the desolate, cooling landscape. The waters accumulated within the bowl of newly erupted lavas. A lake was formed, surrounded by distant volcanic highlands. The wetness within the bowl rose and fell and deposited its history in the fallout of sediment that gradually filled the basin. Hominid toolmakers and other mammals came and went, leaving behind broken remnants of their presence.

Today, you see a desiccated surface where wind and water have pressed erosional fingers into deep ravines and along shallow, rounded slopes. These exposures are largely barren of vegetation. Their fertility lies in the richness of stone tools, fossilized bones, and the detritus of past environments they have liberated from the grasp of strata.

The eroded gullies reveal various sedimentary hues of brown, gray, and white. Our excavations near the base of the pile of sediments illustrates the meaning of these colors. Throughout the region, the terrain possesses a glare of sun-intensified white—the old lake beds. Each layer of white contains billions of microscopic skeletons of algae, the diatoms I referred to in Chapter I. These organisms live and die in lakes, and at Olorgesailie the particular kinds of diatoms and the extent and purity of the diatomite strata tell the history of the lake—its variations in size and chemistry over the past 1 million years.

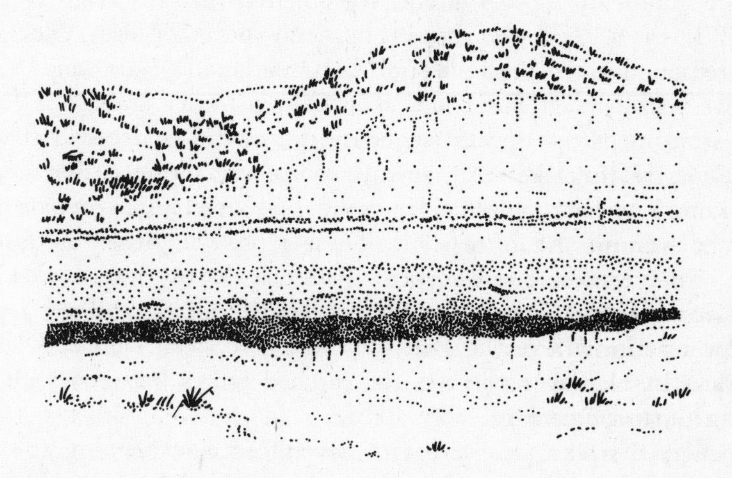

In the oldest chapter of the Olorgesailie record, sandwiched between two huge white deposits, is an undulating layer of dark brown. This is an old soil, pocked where the now-decayed roots of an early Pleistocene flora had penetrated the ground. The plants had colonized a drying mudflat from which the lake had recently retreated. The time was about

990,000 years ago. Numerous zebras, gazelles, monkeys, rhinoceroses, hyenas, and hominid toolmakers infiltrated the area. For several hundred years, certainly no more than a thousand, the terrain was inhabited by these dryland species. Sharp stone flakes were littered over the ground, pressed into the turf and entombed during centuries of rainy seasons. Decaying bones and durable animal teeth were buried by the same process, leading to their preservation.

Within the pile of sediments, about four meters below the soil, a thin continuous streak of gray interrupts the white lake beds. This layer signals an old volcanic eruption. Rich in ash and pumice that once rained over the old landscape, such layers punctuate the geologic record in the Olorgesailie region. They contain radioactive elements that decay at a steady and predictable rate. A method called argon laser fusion has been applied by my colleague Alan Deino to date individual volcanic crystals within the thin gray streak below the brown soil. The crystal ages measure about 992,000 years old. Higher up, fourteen meters above the soil, another volcanic layer can be seen. Its age is 974,000 years.

The intercalation of gray, brown, and white allows us to determine the environmental legacy of this region, from broad sweep to minute detail. In the 18,000-year interval between volcanic pulses, Olorgesailie endured an agitated push-and-pull between lake and dry land. Studied by two members of our team, Kay Behrensmeyer and Tom Jorstad, each stratum in the upward sequence marks the landscape's changing mood—a shift in lake level, an invasion of vegetation, a divergence of a stream. The lake beds display minuscule alternations between pure white diatomite and brown silt. These represent pulses of terrestrial mud into the lake. The lake bed's pallor is sharply interrupted by an overlying brown soil; the lake made a lengthy retreat and was replaced by dry terrain. Above the soil, the rivalry between land and lake intensified. A tan layer portrays the pitched forays between white lake sediments and brown soil formation.

Finally, the lake returned with a vengeance, occupying the entire basin. Its influence spread farther than in any previous era. Its waters lapped against Mount Olorgesailie and licked the surrounding volcanic highlands in the distance. A cursory scan of these deposits suggests a stable, uninterrupted wetness—an eight-meter thickness of white diatoms. But stepping closer, you see that a bold white line disturbs the sequence. The soft, surrounding layers of diatoms are eroded away, leaving a hard, bleached ledge a meter above the brown soil, in the midst of the lake beds. It is visible for over two miles.

This bold white line, a mere ten centimeters thick, is, to me, a profound symbol of environmental change. It consists of calcium carbonate and contains crystals of the mineral halite, which indicates high rates of evaporation. The bold white line is akin to a desert salt flat, and it lies in the midst of sediments indicating a large fresh water lake. We don't know how long it took the lake to dry out, or how long the salt flat occupied the landscape. But the lake did dry up, entirely, as far as we can see. The rise of the lake was halted by ruthless drought; the lake waters then returned with equal abruptness.

The larger chronicle of Olorgesailie is told in Chapter V, which explores the last million years of the human saga. For now, we may note that as time went on, the lake-land fluctuations continued. Then, in an act of finality, sections of the rift floor rose up and became walls, while earthquakes—what geologists call tectonic forces—proved the awful brittleness of set lava. The hardened trachyte bedrock fractured and rose up at the faults. The battle of Olorgesailie, between lake and land, was decided. Earth movements permanently emptied the lake and elevated the terrain, making the million-year biography of the region available to erosion. In a succession of sudden acts over the last 300,000 years, the lava floor throughout the southern Kenya rift was broken and displaced. A checkerboard of cracks, upthrusts and downthrusts, developed for miles around.

In common parlance, the word "fault" would seem to reflect a simple accident, a geologic faux pas; a mere slipup. To early hominids who lived in the region, such movements of lava bedrock were nothing less than catastrophic. And now all we see are dark-brown walls of stone, the edges of lake basins that will never again hold the waters of lakes.

Is the history of this area an anomaly? This first glimpse into the long environmental record of Olorgesailie rearoused my ponderings in the "place of thieves": How pervasive were alterations in the habitats of human ancestry? To what extent and at what pace did nature conduct its own face-lift of the terrain? I have also come to wonder how the pulse of change might have affected the process of natural selection, and how this process, in turn, would have paved the evolutionary paths of bipeds, from which a species would later arise to impose its own rude turbulence on Earth.

2

If we ascribed awareness where none existed, we would discern the climates of past eras as remarkably deft and devious in leaving behind signs of their existence.

Earth's history of temperature and ice, moisture and desert dust, is found in things shockingly small, or in fathoms distressingly deep. Minuscule grains of pollen, pieces of fossil leaf, tiny shells from the ocean bottom, rows of ancient teeth—the depths of Earth hide what at first seems a cacophony of clues about the time in which humans evolved. Yet each line of evidence offers a faint tone, a kind of analytical phonic, which scientists strive to audit, a signal from the past.

The record of climate is constructed by unearthing fossil-rich strata on land and by piecing together long sequences from ocean cores. These archives demonstrate that over the past 50 million years, Earth has cooled, aridity has spread, and the ground has become host to a thinning vegetation. These trends in our planet's history have extended to the present and become exaggerated over the past 1 million years.

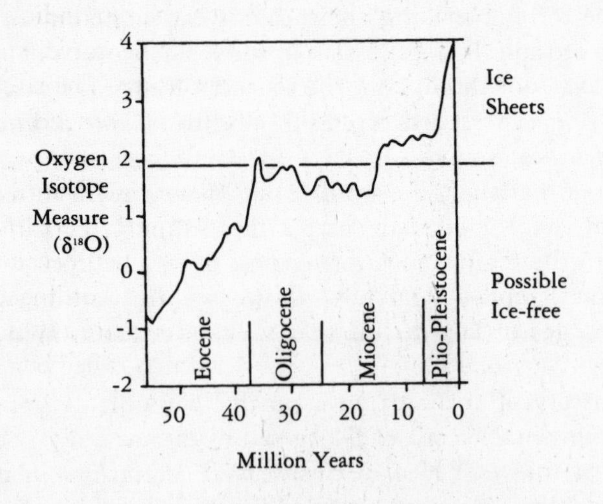

The jagged line above conveys the global schedule of ice buildup and temperature change over the past 50 million years. Its upward course depicts the chilling of the planet and the growth of glaciers.

It is extraordinary to me that such vital signs of the ancient past can be drawn. In this case, the line comes from painstaking measurements of oxygen fixed in the skeletons of single-celled organisms, the foraminifera, or *forams* for short. The remains of these microscopic beasts are abundant in long cores drilled out of the ocean depths. Frozen water, like its liquid form, contains an atom of oxygen for every two of hydrogen. But the lighter isotope of oxygen, ^{16}O, is more easily evaporated from the oceans and incorporated in glacial ice than is the heavier form, ^{18}O. As a result, the oceans fluctuate in their oxygen isotope content depending on how much ice has formed worldwide. As the lighter oxygen molecules are subtracted from the oceans and held captive by glaciers, the ocean-dwelling forams become burdened by heavy oxygen. As ice melts, the lighter isotope is returned to the oceans, and the foram skeletons indicate that this has happened.

Regardless of ice buildup, temperature also affects how easily the two kinds of oxygen are absorbed and recorded for posterity by the sensitive protozoa. And so the long, continuous burial of forams in the seafloor gives us an oxygen isotope curve, both a global thermometer and a record of ice volume. The scale on the diagram's left shows a way of measuring the oxygen isotopes in forams, measured in parts per thousand.

The horizontal bar in the diagram marks the oxygen value at which sheets of ice grew on the continents. Below the bar indicates an Earth that was essentially free of ice; in this part of the curve, an increase in the isotope measure means a drop in temperature. The slightly jagged line from 55 to 35 million years ago depicts an astounding plunge in global ocean temperature, even before there was any significant ice buildup. Above the bar, the zigzag course suggests the overwhelming influence of ice on the oxygen values of the great oceans. The dual process of overall cooling and icing of the planet became prevalent about 15 million years ago. From that time on, according to the mysterious little forams, low global temperatures have been typical of Earth; and about 3 million years ago, a massive volume of ice began to build.

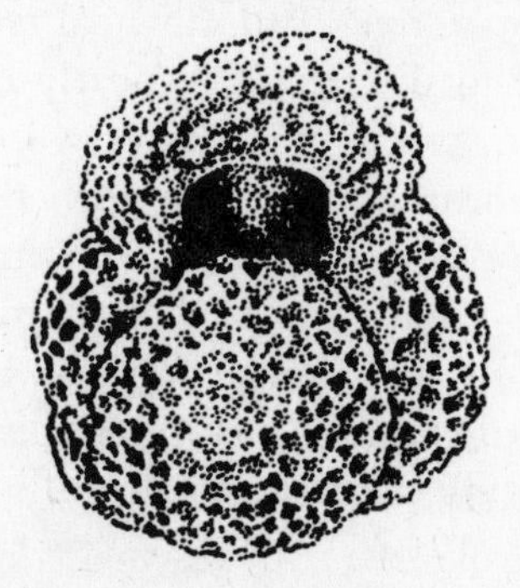

The residues of oxygen deposited in the ocean bottom may seem like complicated science, but the chemical signals left behind by the tiny forams are a kind of poem, an allegory of two grand themes in the study of our planet's history.

The first is: *Earth is a fantastic tangle of connections.* As illustrated so well by the minuscule forams, extraordinary linkages exist between Earth's physical mechanics and its various instruments of life support. The temperature of the planet determines the growth of ice sheets, a process that draws water from the sea. When locked up in glaciers, the amount of water in the oceans and atmosphere inevitably declines. Sea, air, and continents exchange water by cycles of evaporation, circulation of winds around the globe, rainfall, and the rivers' refund of water to the oceans. An enormous plumbing project encircles the globe. And it is linked to an even larger heating system whereby Earth's position in its orbit around the sun, the total amount of solar radiation, the global distribution of heat, and its movement by winds are all intertwined. Air, water, and the dry particles that comprise the land crisscross in a vast network, a complex switchboard of direct connections over immense distances. Modeling the climate is big business in science these days, and one thing amply demonstrated by this endeavor is that polar ice caps, the surface temperature of the sea, the climate of the tropical latitudes, the upwelling of coastal waters, the rise of mountains, and the directions of winds—these separate and seemingly unrelated aspects of the planet—all form a single, intimate, highly connected system.

This has a vital place in our pursuit. The climatic records of numerous far-flung locations help us to envision the environmental conditions of the planet during the course of our evolutionary origin. At every prehistoric turn, we must dig to ascertain the ancient habitats at the very places where our forebears lived, and the conditions that impinged directly on their survival. But our goal requires us to see beyond the limits of particular fossil sites. The nearly continuous climatic records of the deep ocean, where the forams wrote their miraculous archive, are connected beneath the surface with the local settings of our origin. On land, the uplift of plateaus and mountains as distant as eastern Asia altered the circulation of air and moisture in African longitudes where the oldest hominids left their bony traces. The fossil deposits of remote continents also testify to the factors that influenced the evolution of life during the last several million years while in Africa an upright, brainy ape was being born, whose descendants would occupy the planet and make it their own.

The first grand theme is the worldwide connection of environmental systems—and the evidence of them that we discover. The second theme, allegorized by the simple foram, is this: *Earth is a place of peculiar mirrors,* a place where one thing is reflected or recorded in something else, often in odd and unexpected ways. In building its skeleton, the foram reliably records the oxygen content in the ocean at any given moment. The ocean oxygen changes in response to temperature and the buildup of ice on a global scale. With its multitude buried in the deep sea, the humble foram is a shocking mirror of this enormous process, a kind of code into the grand history of the planet's plumbing and heating systems.

These signals can, moreover, be found in great diversity. As the ratio of heavy and light oxygen is the registrar of world temperature, so the volume of terrestrial dust drilled from the deep ocean mirrors the history of aridity on land. I'm not sure whether scientists expected to find large amounts of desert dust far out in the ocean, but it seems a surprising thing to me. The quantity of dust buried in the ocean bottom near northwest Africa, for example, tells us about the phases of vegetation, aridity, and moisture in the Sahara desert.

Plants offer their own reflections on the state of the world and its changing status over time. The physical attributes of leaves are an example. Small, jagged-edge leaves, like those found in northern forests, are a sign of a cool and seasonal habitat, whereas large, smooth leaves with narrowing tips indicate a humid tropical climate. From the distribution of fossil leaves, paleontologists have ascertained that 55 million years ago, warm and humid evergreen forests extended widely over the globe, virtually to the poles. The hard, tattered remnants of ancient plants tell of a green planet without swaths of yellow-brown prairie or desert white. Later on, fossil floras from around the world portray a steep decline in tropical forests, which were segregated to the equator as temperate forests, woodlands, and ultimately grasslands began to dominate.

Characteristics of animals are also reliable signs of environmental change. Among the herbivores, for instance, crested, high-crowned teeth, as seen in horses, evolved over time. They are superb signals of a diet consisting of grass, fiber, and grit, contrasting with the lower, puffier enamel seen in the cheek teeth of animals that eat fruit or tree leaves. The vegetation history of the world is captured in the teeth of animals who eat vegetation. Those of us who wish to feed upon history must harvest these extraordinary signals preserved in fossils and dirt.

Just as the heating and plumbing systems of the world are interconnected, so animals, plants, and physical environments are linked in a network of mutual influence. The fossils and strata of long ago offer wondrous remnants of that network, signaling the environmental history of eons and enabling us to inspect the conditions of the world experienced by human ancestors and other creatures who thrived at the same time.

3

About 55 million years ago, early in the Eocene Epoch, Earth was warm and moist, tropical and homogeneous in its vegetation:

- As you can see in the earlier diagram, the oxygen curve furnished by the forams indicates a distinct high in global temperature at this time.
- Fossil leaves, seeds, and other plant parts recovered from hundreds of sites signal a dramatic expansion in tropical vegetation. Rain forests extended from the equator to 65° latitude, an enormous range compared with 20° around the equator today. The fruits and seeds of the famed London Clay flora, which thrived in England 50 million years ago, indicate a tremendously diverse vegetation, most similar to that of tropical Southeast Asia today. Broad-leaved, evergreen vegetation occurred north of the Arctic Circle.
- The drifting continent of Australia, the isolated landmasses of South America and Africa, and the huge northern areas of Europe, Asia, and North America all possessed tropical rain forests. The consistent fossil record of moist, dense forest suggests the presence of equable climates and vegetation over the face of the planet. The habitats in one place were more or less similar to conditions in any other place. The equable climate of the early Eocene contrasts with the extremely diverse regimes of temperature and rainfall in the present.
- Very similar arrays of animal species are found in fossil sites over each continent. Animal communities appear to have been both homoge-

neous and stable, rather than varied from region to region. This line of evidence supports the idea that early Eocene habitats were strikingly similar despite vast differences in latitude and longitude.

But there was irony in this idyllic world. Although warm, moist, and forested, the planet was poised on the edge of precipitous decline. A downward spiral of temperature and moisture would lead to a world where ice, desert, and steppe would compete with tropical forests for a place on the continents. A vast body of evidence defines these regressive trends:

- The deep-sea temperature at present is estimated to be about 12° to 15°C colder than in the early Eocene. The foram curve takes a startling zag toward the cold end of the thermometer beginning 50 to 55 million years ago. This cold trend became exceptionally pronounced by the early part of the following geologic epoch, the Oligocene.
- Antarctica first experienced major ice buildup during the late Eocene, and by 15 million years ago glacial ice had become a persistent element in the planet's spectrum of habitats. Around 3 million years ago, glaciers began to form in the Northern Hemisphere, and they have swelled and pulsed to a jagged rhythm ever since.
- As Earth cooled, the once-enormous reach of warm, moist forest diminished. The size and shapes of fossil leaves show that tropical rain forests during the early Oligocene were constricted to about 15° latitude around the equator. Temperate forests of deciduous trees expanded in Asia, Europe, and the Americas. Spruce and hemlock, associated with temperate climes, made a brief appearance even in tropical Borneo during the worldwide temperature decline of the Oligocene. Vegetation composed of small, leathery, arid-adapted leaves also originated and eventually spread—particularly in Africa and Australia.
- Fossil animals found together in Mongolia testify to the oldest open-country, grass-dominated habitat known in the Cenozoic Era; it is about 30 million years old. The extensive record of fossil pollen shows that grasses and the family of composite flowers (daisies, asters, and ragweed) first appeared in abundance on the northern continents during the Oligocene, and low-lying vegetation began its invasion of soils around the world.
- By 25 million years ago, the acacia floras, the she-oaks (*Casuarina*),

the juniperlike *Callitris,* and other trees that thrive in droughts were established in Australia—all at the expense of the broad-leaved rain forest. The looming, fragrant eucalyptus woods arose as the climate continued to dry and to cool, so even the richest forests of this southern continent are associated with aridity.

- By 20 million years ago, the rain-forest belt across the midriff of Africa was broken, interrupted by trees, herbs, and grasses that benefited from the gathering aridity. Virtually all the fossil floras of the early Miocene Epoch have species that inhabited dry, open woodland or dry forest.
- Grasses and forest trees left their chemical signatures in ancient soils; analysis of the soils shows that grasslands had become a persistent habitat in Asia by 7 million years ago and formed huge tracts over the past 5 million years on all of the vegetated continents.
- The Sahara Desert was clearly in place by 2.5 to 3 million years ago. While other deserts may be less than 2 million years old, coarse windblown sands, indicating aridity in northwestern Africa, have been discovered in ocean sediments of three earlier periods, as far back as the early Oligocene.
- During the Cenozoic, large-bodied mammals began to develop specialized teeth and legs that were effective in grazing and moving widely and rapidly in open terrain. Over the past 10 million years, these species have become the dominant large land mammals. We see, for example, the ancestors of the migratory wildebeest and cursorial gazelles of Africa; diversification of hopping kangaroos that graze the Australian outback; and forerunners of horses and zebras, which first arose in North America and spread dramatically over land bridges to Eurasia, Africa, and South America. Major blossoming of these groups of animals—between 14 and 2 million years ago—corresponded to dramatic declines in temperature, moisture, and forest foliage over the continents.

This outline is the headline news of our current geologic time, the Cenozoic, the era of new life. A truly voluminous body of mirrors and clues demonstrates the same ponderous conclusion: Over the past 50 million years, the climate and habitats of Earth have deteriorated dramatically. The warmth, forests, relative stability, and homogeneity of the Eocene are a far cry from the quixotic, fluctuating world of today—or even that of, say, 1 million years ago, well before the human lineage harbored any peculiar ken for altering its environment.

It would be gross miscalculation to consider these distant stirrings of minuscule importance to the present. Our planet as it now exists was fueled by the history of cooling, drying, and associated change in plants and animals. That history was prelude to the human story, and its enduring stamp on our investigation requires us to move in for a closer look at the large animals that evolved during the decline.

4

The past 50 million years appear to have been a time of progressive deterioration. On every vegetated continent, grasses and small hard leaves widened their domain. While certain herbivores accommodated to the new foliage—and evolved the art of eating coarse, fibrous fodder—other species, committed to Eocene and Oligocene ways, died. In many animals, from horses and zebras to rodents, from antelopes to kangaroos, from giraffes to camels, from elephants to pigs, the main food-processing part of the jaw, the cheek teeth, became larger in overall height and riddled with complex slicing edges. An increase in tooth height, called hypsodonty (high tooth) is typical of animals that harvest greenery close to the ground—mainly fibrous grasses in open territory which is generally poor in nutritional value.

Any animal committed to such a diet must eat an enormous quantity. This requires both a large gut and a large body capable of handling masses of grass and other fibrous plants. Long-distance travel allowed access to such volumes of vegetation, mowed and chewed as the animal walked along. After the Eocene, many lineages of animals evolved to a large size, formed social herds, and modified their limb bones in ways that allowed wide and rapid movements over open terrain.

This mixture of traits became a kind of syndrome—a complex of characteristics that sprang up on every continent, including those isolated from other major landmasses. In Asia, Africa, and Europe, a superdiverse fauna of Miocene antelopes, horses, giraffes, rhinoceroses, elephants, and their predators—the cats and hyenas—stretched across a vast area. Large, wide-ranging grazers and browsers thrived in an increasingly open swath of woodland, grassland, and forest. This fauna maintained its integrity over a long period of time, between 10 and 5 million years ago, with little change in its overall makeup. This diverse assemblage of open woodland animals is known as a *chronofauna,* literally "a fauna through time."

The open-country chronofauna of Eurasia is not the only one to have evolved during the Miocene. A strikingly similar phenomenon can be detected in the fossil assemblages of North America between 15 and 8 million years ago—a chronofauna of open-country grazers and browsers. Like its counterpart in the Old World, this North American chronofauna far exceeded the diversity of species in the rich savanna faunas of East Africa in historic times.

The diversity can be ascribed to a widening mosaic of plants. In Eurasia, according to Ray Bernor of Howard University, large, plant-eating animals were able to feed in three strata—trees, bushes, and ground cover—whose productivity was not limited by either winter frosts or seasonal drought. David Webb, a paleontologist at the University of Florida, ascribes the Miocene biodiversity of North America to a widespread, relatively stable mosaic of open vegetation. In both chronofaunas, a multitude of large-bodied, mobile herbivores fed within a maze of habitats that, despite seasonal changes, were productive year-round. A diverse community of social predators and bone-crunching scavengers accompanied the herbivores.

Between 8 and 4 million years ago, the rich Miocene chronofauna of North America was dismantled. A biodiversity pinnacle of nearly fifty mammalian genera was reduced to about twenty. Webb attributes this to overall cooling, drying, and the loss of equable climates. The immense diversity of the large mammal chronofauna also began to falter in Eurasia about 5 million years ago; a series of extinctions and origins led to a net decrease in the overall number of species. The lineages that survived in both North America and Eurasia tended to be those with high-crowned teeth and legs built for long-distance movement. Grazers with large foraging ranges benefited as immense prairies and steppes stretched across the northern continents for the first time.

The fact that similar chronofaunas and histories occurred in North America and Eurasia during the Miocene, at great distances and with completely different lineups of species, alerts us to the amazing connections between the separate biotas of the world. Although each region possessed its own signature, the similarities have been global in scope.

The southern continents reveal the same theme, with a slightly different rhythm. In Africa south of the Sahara, new forms of antelopes, other ruminants, elephants, hippos, carnivores, and ground-dwelling primates evolved. Caches of fossils from the late Miocene are sparse, but what is available hints at a diversity explosion in sub-Saharan Africa later than in northern zones—probably between 6 and 4 million years ago. In other words, the "Miocene syndrome" had its greatest blossoming in southern and eastern Africa at the very end of the Miocene and during the succeeding Pliocene. Over time, the large mammal fauna of Africa began to overshadow that of Eurasia, as the latter fell from its Miocene peak.

Similar trends left their mark in Australia and South America—immense landmasses completely isolated from other continents during most of the Cenozoic Era. Australia also witnessed the rise of large-bodied grazers, indicative of the cooling, drying, and opening up of the vegetation. Many species of kangaroo appear to have evolved during the Pliocene.

In South America, the diversity patterns are less easy to interpret. The familiar syndrome is evident nonetheless; the major tribes of ungulates developed high-crowned teeth and running legs. A rich array of species was maintained throughout the Miocene and Pliocene.

With the rise of the Isthmus of Panama, 3 million years ago, a tremendous influx of North American animals wreaked havoc on their southern counterparts. Large mammals native to South America declined. As grasslands and semi-arid conditions encroached on portions of Central and South America, nearly two dozen genera of large mammals from the north found a place in southern habitats. The long independent signals of climatic drying and cooling upon these two separated landmasses were finally joined into one.

These continental histories have serious implications for our scrutiny of human origin. Their first lesson is that strikingly similar trajectories of evolution can be manifested independently by distinct groups of organisms, by communities of animals and plants separated by long distances, and even by the biotas of completely isolated continents. These similarities, which we will call *parallelisms,* emerge at the same time that environmental trends and events of enormous scope tighten their grip on multiple regions and groups of organisms. Diverse plants and animals encounter the predicament of how to deal with changes that their forebears never knew. Sharing the conditions of existence—common experiences in similar sequences of environments—is the basis of parallelism. The occurrence of common history has significant bearing on our central problem—explaining *specific* examples of evolution, especially that venture known as humankind.

A second lesson is equally important: Climates and habitats over the past 50 million years have succumbed to a certain tendency, a global deterioration. Grasses originated and spread during the decline of warmth and moisture. Gradually, Earth's vegetation structure was altered. Continuous tracts of open woodlands, grasslands, and ultimately deserts arose. Animals that could accommodate to increasingly arid, savanna environments evolved and persisted. Caught in this crescendo, the first two-legged apes and first practitioners of stone technology evolved in Africa.

The implications return us to home. Everything in our study of the Cenozoic seems to support the savanna hypothesis—the view, urged time and again by the scientific community, that humanity was conceived in the midst of directional change, cooling, drying, and the swelling of grasslands at the expense of original forests. This progressive change in habitat and the evolutionary response of organisms were a kind of environmental forcing. The shift to the savanna, whether gradual and continuous or pulsed and steplike in its development, was responsible for the major events in the early evolution of our kind.

The African forest began to wither. Humankind's complex adaptations were reactions to terrestrial habitats and the dilemmas of the open savanna. That this ramp of decline actually started in the Eocene Epoch seems to testify to a running start, and to similar shifts from forest to grassland beyond the African continent. It seems to be the old, familiar story—the loss of Eden, the venture out of the forest. But an intriguing twist can be found in this version. If the Eocene was Eden—the forest of abundance far removed from our present world—our predecessors were not expelled from Paradise; the forest itself departed from our distant ancestors, leaving them stuck in the eastern and southern zones of the continent.

The ancient story is deeply engraved in the Western mind, and the science of anthropology has offered its own translation, which stands as our tradition: In the savanna, in the challenges of a new habitat, humanness evolved. The directional slide from forest to savanna created the original predicament—an ape without trees.

The solutions offered by evolutionary change—two-legged walking, toolmaking, intelligence, new social arrangements, culture—forged the characteristics of our species. Since major features in the evolution of other animals—high-crowned teeth and adaptations to open terrain—were also responses to the progressive Cenozoic decline, it would seem that the case is closed. There is no apparent need for any alternative explanation of human evolution.

But is this the true story of our origin?

5

Forest vegetation, wetness, and heat stretched generously over early Eocene landscapes, from equator to poles. But according to the vast body of details in our planet's biography, this state of affairs eventually succumbed to zoning laws. Pieces of tropical forest were parceled and confined as global forces gerrymandered and rearranged habitats. Directional cooling, drying, and the spread of grasslands seem to have been the main thrust of the late Cenozoic. But an equally important trend has been the replacement of sameness by intricate variation. Eocene homogeneity gave way to well-defined zones and provinces.

A deep current of cold water now surrounds Antarctica, and information from ocean cores suggests that it developed around 20 million years ago. Today, the current guards the coast of the enormous Antarctic ice fields and buffers any marine or atmospheric current from infiltrating and landing even a minor battalion of heat from the lower latitudes. It is believed that this current was critical in the growth of perpetual ice on the continent even during the Miocene. A hot equator was companioned by a frigid South Pole. A steep gradient of heat linked the two regions, and the distinction between warm and cold times of the year—annual seasons—became larger all over the world.

As the gradient of heat and moisture developed, the growing seasons of plants had to follow. And as the seasons of the year became more marked, so did the diversity and the zoning of vegetation. Tremendous sweeps of tropical forest were replaced by more open vegetation—but this is the only change in plant life that could have occurred as climate diversified and global zones became better defined. The directional trends in animal and plant evolution were promoted by the mounting *heterogeneity* of conditions that organisms faced.

Heterogeneity actually took two forms. First, distinctive types of habitat were divided and joined in a patchwork or mosaic over the landscape. I imagine the forested Eocene terra, and the foods and other resources available on it, to have been like an enormous, patterned carpet. Over time, the carpet became tattered and replaced by numerous area rugs of smaller and smaller size, which could slip over the global landscape far more rapidly as the complex switches and chains and levers that control climate turned off and on.

A recent study, citing the work of paleontologist Jack Wolfe, summarized this trend: "The vegetational zonation seen today, ranging from tundra and taiga at high latitudes to tropical forest at the Equator, and with a significant portion of the tropical and subtropical land mass consisting of desert, probably represents a more heterogeneous global vegetation than at any other time in tetrapod history," that is, in the past 350 million years of four-legged animals.

The second type of heterogeneity was expressed in the fourth dimension. Over time, environmental swings were of larger and larger magnitude. If, as generally assumed, progressive change is the overlying environmental theme in mankind's genesis, we must hear its strong, capricious counterpoint—an underlying, oscillating rhythm. Climatic instability—alternating seizures of warm and cold, moist and dry—has been unremitting in its effect on the grand trends of the late Cenozoic. The time scale of fluctuation has ranged from annual seasons to 100,000-year extremes. The range of environmental oscillation has itself changed at various times. The degree of fluctuation worsened around the time that bipedal apes emerged on the African continent. And from about 3 million years onward, the range of flux increased until, about 700,000 years ago, regular and intense alterations in climate and landscape began to assert a powerful influence.

Fifty million years, or even the 5-million-year span of the hominids, is far too long to maintain any single direction of environmental change, whether we speak of individual regions, entire continents, or the globe. Tectonic outbursts, volcanic spasms, vacillation in the volume of polar ice, opening and closing of seaways, even the irregular orbit of Earth itself were the instruments of environmental variability. Because of them, habitats became more diverse and less secure as the planet cooled and dried. Our purpose here is to uncover the significance of this growing uncertainty.

6

If, every once in a while over the eons, an enormous hand could have reached out to touch and palpate our planet, it would have felt odd new accentuations developing on its surface. Earth, bruised and scarred, was gaining character. Certain areas swelled, while others had gashes that deepened with time. Earth's face took on a kind of ruggedness. Valleys were exalted, and the plains were made rough places. All of this happened over some 20 million years. As a result, the globe today is a far more mountainous, textured, and interesting place than it used to be.

The enormous plates that comprise Earth's crusty exterior were responsible for much of this character building. These plates move, swell up, and fracture because of the weak, easily deformed rocks beneath them. Nearly 50 million years ago, circulation patterns in these lower, liquidlike masses pushed the plate crowned by what is now the peninsula of India into the Asian mainland. And it kept on pushing at the dull but inexorable pace required by the molten rocks below. The slow-motion collision between the two plates crumpled the crust, creating more than a mile of mountainous relief over the last 10 million years, including the awesome peaks of Himalaya and the massive plateau of Tibet.

The Great Rift of East Africa was also fractured into being. The hard external surface of this wondrous region was pushed up from beneath, causing massive sheets of rock to break and slip along lines that form the rift borders. The formation of faults in southern Kenya, where Olorgesailie and Lainyamok lie, is a microcosm of this activity. The whole thing was well under way by 20 million years ago, as rift faults deepened and volcanic highlands arose on the borders. Olorgesailie and surrounding areas indicate that the rifting process has continued to alter the landscapes of East Africa well into recent geologic times.

Forty million years ago, the High Plains of the United States, the Rocky Mountains, the Colorado Plateau, the Basin and Range, and the Sierra Nevada were all either shallow oceans or low terrain. Pressures in the Earth caused tremendous faults and unprecedented uplift in this vast region of North America. Much of the action occurred over the past 10 million years.

It is no miracle that events fixed to the ground have inordinate influence over the weather. Our planet's surface and atmosphere have long enjoyed constant intercourse. In the time period we are investigating, huge regions experienced faults, uplift and depression, which altered climates in four major ways:

Rain-shadow effects: Because air, when it rises and cools, cannot hold as much water vapor, rain clouds became depleted in the cooled upland terrain, causing aridity to increase on the lee—in the rain shadow. From Miocene to Pleistocene times in eastern Africa, volcanic terrain emerged above the deepening rift valley and intercepted the eastward flow of precipitation across the continent. Western and central Africa continued to receive its abundance of rain, while the rift valley beyond the central plateau became susceptible to drought.

Buildup of heat and release of moisture: High plateaus are easily heated by the summer sun. And so, as uplift proceeded, the temperature difference between ocean and land increased, and the resulting swirl of air currents and the seasonal movement of moisture began to produce monsoons. These rains are especially well known in Asia along the southeastern edge of the Tibetan plateau.

Effects on distant air currents: The huge plateaus' combination of summer heating and winter cooling causes air to sink and rise in the vast surrounding region. According to computer simulations of the process, the awakening of highlands in Tibet led to intense summer drying in the subtropical latitudes of Africa and western Asia.

Increased weathering of highlands: The more relief in the landscape, the greater the erosional effects of rain and wind. Carbon dioxide in the atmosphere interacts with rocks to form bicarbonate, which is washed into the sea by erosion. Thus, as newly risen uplands are eroded, carbon dioxide is subtracted from the air. This lessens the greenhouse effect and is a factor in global cooling. Crumpling and gashing of Earth's crust over the past 10 million years, and subsequent erosion, has significantly lowered the amount of carbon dioxide in the atmosphere.

In some places, the land was elevated perhaps half a mile per million years, creating stunning effects near and far away from the source. This enhanced the growing division of climatic zones, seasonality in rainfall, and provincial grouping of vegetation and faunas. Patchiness and heterogeneity were signs of the time.

The effects were dramatically amplified by volcanic eruptions, another cause of environmental change over the past 20 million years. According to paleontologist Ralph Taggart and his colleagues, spasms of volcanic activity ruled the Miocene landscape in the Pacific Northwest of the United States. Dense broad-leaved and conifer forests were repeatedly laid low by tremendous blasts of volcanic ash, gas, and associated mudflows. After the forests were destroyed, the dry, desolate landscape was repopulated by forbs and grasses. The succession to full forest vegetation took centuries to complete, until the next major eruption started the process all over again.

Seizures of volcanic material on the landscape repeatedly transform it, sometimes over enormous areas. Fitful fluxes alter the conditions of plant and animal survival for long periods of time. Populations that have evolved ways of adjusting to drastic disturbances move into the area and disperse as the pace and degree of environmental disruptions become magnified.

In North America, uplift and rain shadows gradually enhanced the aridity of the region; as this occurred, grasses and herbs, which had been merely the first invaders of the volcanically disturbed terrain, ultimately became the dominant plants. While the spread of grasslands appears to be a progressive trend, low, fibrous vegetation got its original foothold from intermittent disturbance and variability rather than a purely directional change in the weather.

Similar tumult was occurring at about the same time in East Africa. Over the past 24 million years, gray layers of volcanic ash testify to astonishing geologic activity in the region. The fractured walls and floor of the Great Rift System are composed of thick, hardened outpourings of volcanic fissures and cones. Geologists have mapped extensive flows of lava and rock debris in the Miocene deposits of Kenya and Uganda. These show that periodic eruptions disturbed vast areas, acting as an Etch-A-Sketch on local and regional landscapes.

At Fort Ternan, a 14-million-year-old site in western Kenya, the chemical makeup of ancient soils preserving the *Kenyapithecus* ape signal the presence of woodland and forest. Scrutiny of fossil bones from the same locality suggests that numerous antelopes moved about in places of relatively closed vegetation. Curiously, paleontologists have debated the correct environmental interpretation of this site for two decades. Fossilized pollen is also preserved at Fort Ternan, and 54 percent of these microscopic grains are from ancient grasses. Other faunal analyses have suggested that somewhat open vegetation within a seasonally dry climate occurred at this site.

Similar discrepancies, or what might appear to be such, arise in the Miocene fossil record of the Tugen Hills in central Kenya. According to various authorities, the fossil mammals found there made their homes in open woodland and savanna habitats. Yet within the same geologic bed (the Ngorora Formation), a beautifully preserved array of 12-million-year-old fossilized leaves has been discovered that suggests lowland rain forests similar to those of western Africa today.

To resolve such apparent conflicts we must recall the moral of Taggart's model: The confusion about Fort Ternan and other Miocene sites of East Africa may lie not with the percentage of grass versus forest but with the assumption that only one true habitat existed at any given Miocene site. Woodland, broader grassy zones, and forests may all have occurred at different periods, and the varied lines of evidence may signal a blend of diverse environments over time within the fossil record.

The crossfire of volcanism, rifting, and uplift makes it difficult to portray the Miocene as anything but tumultuous from an environmental standpoint. Each of the factors fomented cooling, drying, and insult on the forests, and so played an instrumental role in the gathering decline. Simultaneously, each factor promoted the subdivision of climate and vegetation. Especially in cases of volcanism, environmental disruptions were irregular in frequency and extent. As a result, habitats varied in a complex pattern.

Dissolution of the forests during the Cenozoic Era was a process of fragmentation. Continuous tracts were broken, and the boundaries between forest, woodland, and grass became prominent features of the landscape. The drying trend of the same era was caused not only by decreasing rainfall, but by its annual fragmentation and confinement to certain seasons. Even the weather was divided, parceled out. As the topography of Africa, Eurasia, and the Americas became more complex, the air-circulation systems that introduced moisture to large regions were changed. In Africa, a broad band of moist air, which had once cut all the way across the continent, was transformed into a more restricted cellular pattern of circulation sometime between 15 and 5 million years ago. This new meterological setup was later intensified as colder temperatures developed in the Northern Hemisphere. As a result, seasonal differences in rainfall increased, which meant that dry seasons became drier, and monsoons wetter.

Vegetation had to adopt to the seasons of minimum rainfall; woodlands and grasslands insinuated themselves, creating cracks that separated patches of forests. The cracks grew as precipitation was parceled into extreme periods of the year. Once again, heterogeneity increased. East Africa, where fossils of the earliest apes and most ancient humans have been found, was especially susceptible to such unexpected influences.

They were unexpected simply because the ongoing change in terrain and conditions of survival—climate, seasonality, vegetation, resource spacing—had nothing to do with the animal populations involved. Species confronting this dynamic situation were neither primary causes nor controlling agents. The intimate relations between tectonic change and climate, volcanic eruptions and vegetation, lay outside the bounds of animal livelihoods and survival strategies. Even though Earth's uplift and rifting were progressive, directional processes, their effects fragmented the earlier homogeneity of habitats even farther and increased the uncertainty of maintaining life in a particular time and place. Considering the million-year perspective in which lineages originate, endure, and become extinct, the whole situation of mid-Miocene Africa was full of the unexpected.

7

The continent at the very bottom of the globe, now entombed by ice, has had surprising influence over the climates of distant landmasses, both tropical and temperate. As read from the foram thermometer and other evidence, Antarctica began to accumulate broad sheets of ice between 40 and 35 million years ago. The continental plate of Australia had already separated from Antarctica. The seaway that resulted along the eastern edge of Antarctica, so the speculation goes, provided tremendous amounts of moisture, which was precipitated as snow. As a result, massive glaciers formed.

The initial growth of a polar ice sheet became a benchmark of global change. The worldwide distribution of heat, water, and vegetation, which was more evenly spread in earlier epochs, was significantly altered. The Southern Ocean turned cold, and Earth was soon divided into well-marked temperature zones from the equator to the poles. The breakup of Eocene uniformity foretold the planet's future; the buildup of glaciers in Antarctica was a sign of the bitter ice era to come.

For every story, there is a counter-story; for every body of evidence, an old way of thinking that must be changed. The controversy about Antarctica centers on the timing of the ice accumulation. Fossil pollen, leaves, and wood offer abundant evidence that broad-leaved forests occurred in Antarctica *later than* 30 million years ago. If this is so, some scientists believe that Antarctica must have become icebound much later than 35 million years ago.

This controversy has everything to do with our basic view of nature and environmental change: Why is evidence of the oldest Antarctic glaciers assumed to reflect a progressive, permanent change to icebound conditions?

Distinctly mirrored in the record of sea level, fossil plants, ocean organisms, and glacial sediments is the fact that the temperature of the polar landmass underwent dramatic change. The ice sheets fluctuated greatly in size; in some regions, possibly over much of the continent, large glaciers and forests alternated with one another through time. Without question, this southernmost terrain has become increasingly loaded with ice. Nevertheless, climatic variability, rather than a constant or continual rise in ice volume, is the surprise in Antarctica's history.

The foram thermometer suggests that the southern ice cap took on its present form sometime between 15 and 14 million years ago. Since then, enormous ice sheets, thick moraines, and extensive glacial erosion have been Antarctica's hallmark. The continent has remained glaciated for most of the time since the mid-Miocene.

The controversy continues, however, as ocean diatoms have been discovered in the interior of the continent, dated to about 3 million years. If this date proves correct, an inland sea reached deep into the mainland, and considerable warming and melting must have occurred. The strong but still debated possibility of a significant polar thaw at this time would have meant a sharp reversal in the Cenozoic cooling trend. Fossil evidence of southern beech trees has also been discovered. These plants indicate temperate warmth, yet signs of them have been found interspersed with glacial deposits of the Pliocene Epoch, a finding that helps to build the case that the Antarctic landscape experienced rare fits of vegetation within an otherwise dynamic habitat of ice. Even within the time frame of bipedal humans on Earth, it is possible that this most stable of all continents has felt the hard throb of environmental change.

No factor is more important in grasping the climatic odyssey of Earth than its variable orbit. Our planet's relation to the sun changes over time, which varies the amount of solar radiation hitting its surface and influences world temperatures and ice buildup. Here we can see the global scope of environmental change in full view.

The foram thermometer has many lessons to teach, not least of which is the rhythm of nature's own pulse. Its mirror of world temperature and ice buildup, depicted on page 50, has convinced us of the decline in our planet's climate. Defined by cooling, aridity, and deforestation, this decline is the broad signal of the past 55 million years. But on closer observation we find that the enormous sweep of the foram curve from warm to cold is actually composed of innumerable deviations, many more than can be depicted in our earlier diagram. The apparently smooth line has teeth, the jagged edge of nature's alteration. The oxygen contained in the miniature skeletons of ocean-dwelling forams has undergone tremendous oscillation. The forams absorb heavier, lighter, then heavier forms of oxygen, as the globe swings repeatedly between greater and lesser amounts of ice. From the frequency of these shifts, and from calculations first made by the Serbian mathematician Milutin Milankovitch, it is clear that the pace of these alterations is controlled by eternal variations in Earth's orbit.

Over the course of a year, Earth moves around the sun in an elongated path, an ellipse. It's been known since the 1840s that the elongation of the orbit (the flattening of the ellipse) varies slowly and continuously. The orbital paths of all planets shift predictably, according to their different speeds around the sun and the ever-changing gravitational pulls upon one another. A complex cycle of gravitational pulls ensues, stretching and shrinking the shapes of the orbits over time. For Earth, this cycle of change is approximately 100,000 years. In times when our planet assumes a flattened path, it closely approaches the sun *and* moves far away from it within a given year. As it obtains a more circular orbit, a fairly constant distance from the sun is maintained. The amount of annual fluctuation in sunlight is thus altered on this long cycle.

Over the course of a day, Earth rotates on a tilted axis. The tilt is currently 23.5° (a deviation measured from the plane of Earth's orbit), but this angle fluctuates gradually on a cycle that lasts 41,000 years. As a result, the polar regions' incline toward the sun varies—enough to make the amount of sunlight that embraces these ends of the globe change significantly over the lengthy cycle.

Because of its tilt, our planet has seasons—variations in the hours of sunlight throughout the year. And thanks to the 100,000-year cycle in the orbital ellipse, the seasons come and go with ever-changing distances between Earth and the sun. As if this weren't complex enough, our planet has a wobble in its axis of rotation, which also changes the timing of the seasons relative to the distance from the sun. Interaction between Earth's wobble and its longer orbital variations causes yet another cycle. Although the Northern Hemisphere winter now starts when Earth is very close to the sun, the opposite occurred about 11,000 years ago. At that time, winter began when Earth was farthest from the sun, which meant much colder winters in the Northern Hemisphere. This cycle, known as precession, occurs regardless of whether our planet's orbit is flat or more circular, and it takes place over approximately 23,000 years.

It is obvious that these matters of orbit, tilt, and rotation influence Earth's reception of sunlight and its annual distribution. That these factors also act over cycles of very long duration enables us to comprehend an important fact about the fundamental time frame of climatic variation. Rising and falling over 100,000 years, 41,000 years, and 23,000 years, the threads of change intertwine and their effects of warmth and cold promote and contradict, amplify and moderate each other.

Oxygen measures of foram skeletons spanning the past 1 million years, the period known as the Ice Age, indicate a complicated rhythm of fluctuation. The zigzag pattern of oxygen reflects the growth and decay of vast ice sheets. When this pattern is scrutinized by spectral analysis, the dominant cycle of ice and temperature change is 100,000 years long, corresponding to the expected shift in Earth's orbital shape. The next most prominent pulse occurs every 41,000 years, the predicted variation in Earth's tilt. The next pulse is 23,000 years—the precession cycle.

These astronomical cycles have sometimes been called "the pacemaker of climate change," but the planet has also experienced odd changes in environmental variability. According to the foram curve, the degree of fluctuation in global ice and temperature has varied over geologic time. The 41,000-year cycle dominated during the Miocene and Pliocene, whereas the mighty 100,000-year pulse has ruled over the past million years. No one knows exactly why this shift in the rhythm and strength of environmental change occurred. Earth's orbital cycles obviously asserted great influence, but they represent only one factor. The internal characteristics of the planet—ocean currents, volcanism, tectonics, and regional idiosyncrasies—also create a strong interactive effect. All of the factors combined are responsible for the unique environmental signature of each epoch in Earth's history.

In many years of trying to describe these elements of astronomy, I have never quite overcome my unease about matters that lie so far outside myself. It is sometimes comforting to sense the apathy others have about these unfelt phenomena. But I realize the irony in this. The grand, intimate workings of the planet on which we have evolved are matters we do not think about, do not even consider, because they lie beyond our interests. So we remain deaf to the fantastic rhythms and engaging motifs that manipulated the climates and moved the landscapes in which our present condition developed.

The world is continually changing, and the extremes of sunlight and seasonality, ice ages and interglacials, intense aridity and precipitation, themselves change on cycles of tens of thousands of years. They may be thought to be beyond the scale of our attention, but our capacities to modify the planet, which are evident well within our lifetimes, are somehow bound to the extremes that exceed a lifetime, or a hundred lifetimes. Humanity itself emanates from nature's own complex pulse within our particular era of geologic time.

8

By 20 million years ago an ape known by the name *Proconsul* had evolved in Africa. It thrived in the eastern sector and actually consisted of several different species, as we can see from fossil bones found around Lake Victoria.

The proconsuls were at home in the trees. Their bones articulated in the skeletal arrangement of an agile four-legged climber. But they possessed an odd skeleton, ill-matched to our definitive categories of ape or monkey. In one species the flexible backbone, narrow chest, and part of the pelvis most closely resemble those of a monkey the size of a baboon. Other aspects of its pelvis, however, its lack of a tail, the wide movement of its hip and shoulder, and its enlarged brain all suggest affinities with the apes. It is this essential contradiction to our present fixed boundary between monkey and ape that shakes us into recognizing that evolution has occurred.

The proconsuls lived in dense forests, so it might at first appear that these very early apes were not yet touched by the drying and opening of vegetation, or by the increased variability of environments. But as we have already noted, the East African rift was stirred by unwarned movements of the land and sweeps of volcanic eruptions. The proconsuls originated and diversified in this setting.

The early Miocene apes mirrored the changing world by multiplying, dividing, and spreading in the subsequent period. Between 18 and 12 million years ago, the continental plate that holds Africa and Arabia nudged farther and farther into the plate of Europe and Asia. The broad seaway that had once divided these two gigantic masses of Earth's surface became narrower and narrower. And from the exchange of species between Eurasia and Africa we see the first evidence of a land bridge between the two areas. *Proconsul*'s descendants were among the mammals who found their way into the southern zones of Europe and Asia. During this interval, at least four major groups of apes branched apart.

It was the apes' turn to toss the evolutionary dice, and they flourished in the intricate mosaic of forests and woodlands across the Old World. The significance of this acme in ape family history, 18 to 12 million years ago, cannot be overestimated. Out of this radiation of species arose a certain ancestor, the node from which the lineages of modern apes and the bipeds of Africa would branch. What caused this radiation of apes, this quickening in the labor of our origin?

In our search for clues about the environmental conditions of human origin, let us turn once again to the foram thermometer. The foram curve shows a dramatic angle at the 15- to 14-million-year mark. This emphatic shift, as we have already noted, is a hallmark of the decline, the massive rise in glacial volume, the icing of Antarctica. The rise of Afropith and Kenyapith apes in Africa, and of the Sivapith and Dryopith species in Eurasia, led to the highest diversity of apes ever recorded. This radiation of species coincided roughly with the sharp turn in the foram curve, making it tempting to link the radiation with the cooling and drying of land ecosystems. Was this trend responsible for the origin of the mid-Miocene apes?

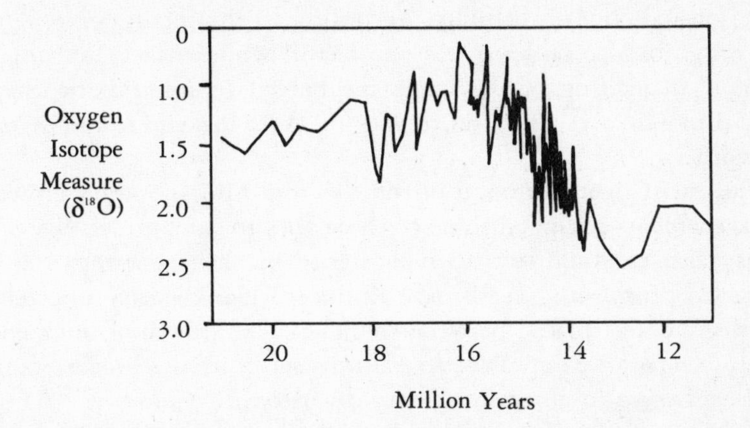

I think we must answer no. The efflorescence of Miocene apes did not correspond exactly to the global temperature decline, as a close look at the foram curve helps to show. According to recent fossil finds, the Afropith apes, which were among the first to possess a thickened cap of enamel on their teeth, arose at least 17 to 18 million years ago—about 3 million years before the frigid dive in the oxygen isotope curve. The oldest Kenyapiths also preceded this environmental episode by at least a million years.

When we consider the foram thermometer between 22 and 12 million years ago, we discover a more intriguing suggestion. Starting nearly 18 million years ago, the amount of fluctuation in the oxygen isotope curve began to increase. Between 18 and 16 million years ago, the average fluctuation was about 0.5 *parts per mil,* the units by which the ^{18}O to ^{16}O ratio is measured. This average amount of change is nearly twice that observed in the prior 4-million-year interval. It is also quite clear that this change in the degree of fluctuation was merely the beginning. It culminated 16 to 14 million years ago in an even larger average fluctuation—up to 0.68 parts per mil—coinciding with a rapid fluttering in the planet's climatic pulse. The Miocene hallmark of habitat decline, the glaciation of Antarctica, thus consisted of numerous large-scale reversals in the cooling and icing trend, not a single major change. The degree and pace of fluctuation first increased about 18 million years ago.

These observations suggest that the rise and early radiation of the mid-Miocene apes—the signal event in primate evolution of this epoch—had more to do with environmental variability than with a directional decline in habitats. According to the foram curve, fluctuation in global ice buildup was one factor, while rifting, faulting, and volcanic blanketing of the landscape helped to revamp the environmental mosaic over time and to vary its exact makeup from one region to another.

The main dental innovation of the mid-Miocene—the growth of thick enamel—equipped some of these apes to eat a wider diversity of foods, both hard and soft, than that tried by their ancestors, the thin-enameled proconsuls. Teeth have limits in their capacity to crack and crush any object placed between the jaws. The growth of thick enamel would have allowed apes the freedom to select from a broader range of foods, ushering in an era of dietary diversity.

If this interpretation is correct, thick enamel in the mouths of the Afropiths and later apes was a kind of insurance policy, enabling them to meet larger uncertainties in habitat and food than their ancestors could have survived. Oscillations over tens of thousands of years proved much more dramatic than any climatic change an ape could have experienced in its lifetime. It is in this longer time frame, as new climatic extremes arrived again and again, that the ape lineages faced their greatest tests of survival.

We have found, then, a middle path, where the real tension between trend and oscillation, direction and uncertainty, can be observed. In the zigzag of the isotope curve from 15 to 14 million years ago, we can see that the key episode of Miocene cooling was actually a fitful throb of change. The rhythm of reversal in the environmental trend suggests that orbital variations were the primary cause. Here we can detect the beautiful coupling of direction and moody complexity in Earth's environmental biography.

We can also begin to discern the entire process by which the survival of species is decided as life on Earth unfolds. Every individual alive today—human or ape, ant or dandelion—belongs to a lineage whose duration is measured in geologic time. Life's vital tension thus consists of, first, the forces of stability and trend that encourage organisms to respond in a certain way according to the conditions of the moment and the predictable future; and, second, the forces of uncertainty, which favor a means of flexibility, not merely the single way that yields maximum benefit in a flinch of time, a season or a life.

The Miocene Epoch created all the conflicts one might imagine could be wrought from this tension. Distinct segregations of species occurred throughout the interval between 20 and 5 million years ago—a phenomenon known as provinciality. Entire zones of vegetation fractured and diversified. The resources they contained, on which the animals relied, became more dispersed, and the patches more delicately defined. Distinct arrays of animals began to be assigned to different regions. This fissioning process fueled the extraordinary diversity of the famed chronofaunas of the Miocene. The yin of stability—the amazing persistence of these faunas over millions of years—was nurtured by the yang of heterogeneity—the breakup of vast zones of vegetation into discrete units.

Direction and uncertainty, the inevitable and the unpredictable, joined hands. Any attempt to grasp the evolution of life, even the singularity of our own origin, must reconcile these two fundamental aspects united in the Miocene. It was an epoch that ended with the advent of our own ancestors, a seed unique to humankind.

The power of continental drift wielded a transforming hand on the major seaways of the Miocene world. By 12 million years ago, Africa had moved forcefully against Asia; the Arabian bridge between the two landmasses had been completed. As a result, the Tethys Sea, which had long provided a continuous circulation of marine waters between the Indian and Atlantic oceans, disappeared, removing a major source of moisture from Africa and Asia. The Mediterranean Sea was defined by the same event.

By this time, the Sivapith apes of Asia and the line leading to the orangutan had already established themselves, and the Dryopiths of Europe were in the midst of their own starburst of doomed species. Shortly thereafter, the record left by the apes began to fog mysteriously. The group of animals we call the Great Apes, perhaps because of their resemblance to ourselves, had already reached and passed its peak; by 8 million years ago, they were in their decline.

We rightly lament the present diminishment of the orangutan, the gorilla, the common chimpanzee, and the bonobo chimpanzee, those living cousins so limited by human encroachment. But, distant as it is, the ape world of the late Miocene deserves our deeper grieving. The Great Apes of that epoch felt even more profoundly the treachery of habitat change. Beginning in the late Miocene, the surviving lineages probably found refuge in hidden, green parts of the world's fragmenting mosaic of forests and woodlands. While the human venture has tested the fortitude of even those refuges, the critical point here is the more rigorous history of decimation, shadowed by time and commanded by nature, millions of years before modern mankind indulged in its own impatient surgery on Earth's habitats.

Extinction of the ancient apes was an omen of a much larger fate. The diversity of savanna-dwelling species in the Old World and North America swelled enormously as the woodland-grassland mosaic marched across the continents. But as the Miocene drew to a close, the chronofaunas of these vast continents unraveled, dying with the species that comprised them. At the very outset of our own evolutionary experiment, with little more than two legs and small canine teeth in our résumé, the large mammals of the world experienced one of the worst biodiversity dives of the Cenozoic.

The balance between variability and stability had apparently gone awry. The isotopic curve and mineralogical evidence from ocean cores support this view. Climatic variations were relatively few and low-level in the interval between 8.8 and 6.4 million years ago. The forces of stability apparently had the upper hand. But not for long. Over the next 1.8 million years, world climate fluctuated more quickly, and oscillations in the oxygen isotope ratio became as large as 1.0 parts per mil—considerably greater than in the unstable period that initiated the ape radiation, 18 and 14 million years ago.

The buildup of glacial ice in Antarctica and around the North Pole became more extensive, removing water, indulging in an assault on the oceans. By at least 6.4 million years ago, glaciations were large enough to isolate the Mediterranean Sea from the oceans. Without a link to the outside, the Mediterranean could not survive the tax levied by evaporation. The sea dried up, completely as far as can be discovered. Enormous salt deposits developed where it had been, and drought infiltrated the surrounding regions.

This long episode, known as the Messinian salinity crisis, created a special looking glass on a newly destabilized world. Painstaking studies of the Mediterranean sediments show that this immense area witnessed not one long drought, but a repetition of intense dryings and fillings. Orbital variations affecting solar radiation and glaciation had an indirect hand at least in this expansion-contraction cycle and in the moody alteration of landscapes on the three continents surrounding the sea. The Messinian crisis thus testifies further to the strong bond between fluctuation and drying, oscillation and cooling, environmental reversal and decline.

The important point is that the degree and pace of climatic alteration—worldwide and in the vicinity of Africa in particular—increased during the period between 6.4 and 4.6 million years ago, coinciding with the Messinian crisis. During this vital span our own lineage came into existence. The earliest hominids were greeted by an erratic dawn when the parameters one usually associates with plenty—warmth and stability—were in doubt. The early humanlike apes took their first bipedal steps during an interval broadly marked by greater seasonal variation, which resulted in drier habitats; by decreased equability in temperatures, which resulted in a cooler planet; and by shifts in biodiversity. The Miocene cornucopia of large mammal species met its demise over wide regions of the world, and in that tempestuous era humanity's lineage arose.

While certain stretches of time since then have witnessed amelioration, the degree and speed of environmental change has largely risen. Born in a time of environmental crisis, the human lineage—or at least the segment that found a way to make things and to alter habitats—eventually became a manufacturer of crisis. It is this culmination that the human species and all organisms living today now face.

9

I have attempted here to sketch my own delving into the environmental history of Earth. That venture began a decade ago in a curious little no-man's-land in the southern Kenya rift. I wondered then, a disease that has not left me, about the singularity of the human lineage. In my pursuit of how the human condition was born, I wondered about the clues of nature's alteration that I saw confined to dirt and rocks, arranged by ancient habitats, stacked and ordered by time.

We have reached deep into the past—far beyond the time represented by Lainyamok or Olorgesailie, well beyond the era of the earliest bipedal ancestors of humankind. And we have greatly exceeded the meaning of a year or a generation, the durations our hearts embrace most fervently.

In that daring stretch of time, there are remarkable things to see, not the least of them the throb of nature. Environments that might be assumed to thrive on equilibrium were instead nurtured on disruption.

Species that might be construed to exist by some consistent, progressive means of honing instead tolerated deep conflict and unforeseeable flux in the parameters of survival. The complex intercourse between animate and inanimate, which might be expected to yield predictable cycles and trends until man's recent appearance, has instead long sired reversals of environmental fortune and daunting precipices of change.

Oscillations between land and water, volcanic outpourings, convulsions of the terrain—these were the gathering events near the first appearance of humanity's unique ancestors. Each element in the saga seems to possess its own reflection in the small pool of ancient lake beds called Olorgesailie. This is our first substantial hint that the dramatic, repeated alteration of the land where our predecessors walked, even within that remote basin of white dust, may not be as peculiar as it first appeared.

CHAPTER IV

EXPERIMENTS IN BEING HUMAN

I

WHAT are the rudiments of human dominion, the oldest beams on which the weight of humanity's environmental dealings now lean?

In the earliest steps of our lineage, the proddings of natural selection began to endow a bipedal ape with human ways. The fossil record has surrendered only the most fragmentary clues near the point where apes and humans diverged. On one side of this crucial divide was a population of ancient apes who eventually walked on two legs and had small canine teeth. These are the oldest elements of the human condition that can be seen in fossils, and of the two, bipedal activity draws our special attention because it changed the way our oldest ancestors moved, setting the groundwork for later experiments in human behavior.

Five-million-year-old molars and jaw fragments found in central Kenya provide tantalizing clues. They appear very similar to those of the oldest definite members of the human family, the early australopiths. But none of the finds gives direct proof of upright walking. Anthropologist Tim White's recent discoveries at Aramis, Ethiopia, show that a primitive species with relatively small canine teeth had evolved by 4.4 million years ago. These fossil remains, known as *Australopithecus ramidus,* narrow in on the key separation between apes and humans, but it is still unclear on which side of the divide they fall. Leg and hip bones this old may ultimately reveal a different way of moving around, or at least a different degree of bipedality, than occurred in the early australopiths.

Finds by paleontologist Meave Leakey in northern Kenya show that two-legged walking evolved by at least 4 million years ago. Leakey's species, *Australopithecus anamensis,* had lower leg bones that carried the entire weight of the body, a clear indicator of bipedal activity.

Much can be learned about the bold world of the first upright, walking hominids by inspecting the rich clues in younger sites. At Laetoli, Tanzania, where Mary Leakey uncovered trails of fossil footprints, volcanic ash and rain fell on an open landscape 3.6 million years ago, and across this terrain three creatures moved erect, on two legs. Like people's, their big toes were drawn in toward the smaller digits, and the inner soles of their feet were arched. Not an isolated print but continuous tracks of two-legged walkers were pressed into a twenty-three-meter length of volcanic muck, now hardened into rock. Fossilized jaws from Laetoli demonstrate that small-canined hominids were active in the area of the footprints.

Other sites 3 to 4 million years old also record signs of early bipeds. The skeleton known as Lucy, made famous by anthropologist Donald Johanson, is among nearly three hundred hominid fossils so far reported from the Hadar area of Ethiopia. These old bones are beautifully preserved and mottled with the hues of minerals that infiltrated and made them hard. You can see holes where nerves once penetrated living tissue, furrows where veins and arteries conducted life through bone, ridges where muscle operated the skeleton, and every other kind of detail known to osteologists. From such patent signals—of shoulders, knees, toes, and virtually every bone in between—a body with distinctive mannerisms and movements is conveyed.

The hipbones of these hominids were broad and short from top to bottom, much more like those of a human than the long blades of an ape's pelvis. Lucy's thighbones were slanted in at the knee, again unlike those of an ape. The weight of Lucy's erect body was thus translated through the midline, as in the bipedal striding of human beings. When chimpanzees and gorillas toddle on two legs, they control their bodily momentum in a very different manner, tossing their weight from side to side, shifting their balance from one leg to the other. Lucy did not walk that way.

Based on the overall relationship of hip to thigh, knee to ankle, and arch to big toe, it appears that the early australopiths secured a new angle on locomotion. They were not tall, but they were upright. The novelty of it all, essentially a redesign of the ape skeleton, suggests a definite and distinctively human commitment to two-legged perambulation and to life on the ground.

That the early australopiths, 3 to 4 million years ago, had already devoted themselves to terrestrial walking on two legs is elegantly championed by anthropologist Owen Lovejoy of Kent State University. He and his colleagues were among the first to address the behavior of Lucy's kind, and they continue to build the case that this critical fulcrum in the development of humanness was forcefully experienced by this ancient species of the human lineage, *Australopithecus afarensis.*

By now, we have little choice but to try to enjoy the fact that scientific pursuits never end so crisply. Just as the general arrangement of Lucy's bones seemed to convince us of a body built for terrestrial walkabouts, bony details were gleaned by other anatomists convinced that Lucy's station in life was largely in the trees. According to Randall Susman and his colleagues at the State University of New York at Stony Brook, the arboreal way and body plan of the apes was not lost to *A. afarensis.* Indeed, it was integral to who they were and how they thrived.

Lucy's toes were longer than ours, though not as extreme and fingerlike as a chimpanzee's. Her feet were crude versions of the rigid paddles on which we stride. Although short from top to bottom, the blade of her broad pelvis was directed more to the side than that in human beings, an apelike orientation that would have provided mechanical advantage in climbing. The ankles of *afarensis* seem to have afforded wider extension of the foot, at least in smaller individuals, than is called upon in the commonplace striding of modern people. The hand was constructed with a powerful grasping ability, and the upper arm was massively built and muscled in a way that suggests an ape's arboreal reach. Numerous other details of arm and leg hint that these founders of the human lineage moved with agility in the trees, climbing into at least the lowest intersections of boughs. The anatomists from Stony Brook thus conclude that the early australopiths climbed, sat, and conducted the main events in their lives from branches above the ground.

But according to Lovejoy and his collaborators, Lucy's pelvis and thighs were positioned in such a way as to provide specialized leverages and momentum for bipedal movement—a two-legged gait that perhaps surpassed that of modern humans. The details suggesting arboreal behaviors are simply holdovers, useless vestiges of a life-style discarded by the early australopiths. In return, the Stony Brook anatomists have laid out the clear, functional clues of apelike arboreality in the australopith body. Anatomists Ron Clarke and Phillip Tobias's recent report on foot bones over 3 million years old supports the idea of an arboreal hominid. According to this report, early australopiths in southern Africa had divergent, mobile big toes that were capable of grasping, even though the foot overall was that of a biped.

For over a decade now, students of human origin have felt the strain between these opposites. Was Lucy terrestrial or mainly tree-dwelling? Did natural selection, guided by an expanding grassland, mold *Australopithecus* for moving about on two legs? Or did the early australopiths cling to arboreal ways, moving bipedally only as part of life in the woods? Was two-legged walking the first human response to treeless surroundings? Or do the scars in Lucy's skeleton register her commitment to climbing trees *and* walking on the ground in a complex but largely forested place?

This clash is caught up in the overwhelming urge to discover the Rubicon of our genesis. That terrestrial life represented such a threshold, with enormous influence over a previously tree-adapted ape, is a pivotal doctrine in the investigation of human origin. The timing of this critical event, the adoption of ground-living ways, is a durable symbol of when humanness began, inextricably tied to the question of what it means to be human from an evolutionary perspective.

According to our orthodoxy, the election of terrestriality over arboreality was accomplished by the process of natural selection as trees became sparse. This is the *savanna hypothesis,* an idea that has nurtured anthropologists for generations: *Adaptation to open terrain was the spark that initiated the human lineage.* Indeed, the fits and starts of environmental decline and the rise of savanna habitat during the late Cenozoic seem to conform to this perspective.

Darwin was one force behind this viewpoint. In *The Descent of Man* (1871), his objective was to draw together the components of this initial act in the human drama. With amazing foresight, he placed the advent of bipedalism, movement into open terrain, and diminishment of the canine teeth side by side. In the loom of human evolution these were the warp, the fundamental strands, woven into the human tapestry. The weft, the thread that bound this primal dressing of humanness together, consisted of tools, the act of making implements. Darwin reasoned that terrestriality in an open habitat implied certain dangers, requiring weapons more effective than long canine teeth. Thus the invention of tools for defensive purposes enabled a protohuman ape to meet the savanna's perilous demands. Manipulating weapons had, however, its own requisites. Primitive arsenals were useful only if hands were free to handle and carry things. Hence, the bipedal imperative. With the origin of toolmaking, the growth of large canines became unnecessary. The energies of life were put instead into brains, cunning, problem solving, dexterity, and new emotional and social contracts. The fabric that is humankind was thus woven together.

It never fails to amaze me that Darwin introduced his scenario with essentially no fossil evidence at hand. Not until the 1870s to 1890s did the Neanderthals become firmly fixed in the human family tree. Not until the turn of the century did a new tide of discoveries, this time in Asia, sweep the more ancient *Homo erectus* into view. Darwin had no inkling of the australopiths, first announced to the world in 1925 and vindicated as true hominids in the 1940s and 1950s. And only during the two decades to follow did advancements in and public curiosity about the fossil evidence of human origin come into full bloom worldwide—much of it connected with the names Leakey, Tobias, Howell, and Johanson. Curiously, it was during this outburst of prehistoric excitement that Darwin's account of human evolution was largely adopted.

The seeds of this movement were sown with less fanfare during the 1950s. Sherwood Washburn and Kenneth Oakley, arguably the leading paleoanthropologists of that period in the United States and Great Britain, took up Darwin's tools as the primary cause of human origin. In 1960, Washburn wrote: "Tools, hunting, fire, complex social life, speech, the human way and the brain evolved together to produce ancient man of the genus *Homo.*" The handling of implements was both a cause and an effect of bipedal movement. Small canine and incisor teeth were "biological symbols of a changed way of life." The protection of the social group shifted from teeth to tools. Carrying, manipulation, dietary change, social life, culture, and two-footed gait were all tied together in an inevitable, closely knit package. Later, fine tuning of the human adaptation largely involved the brain, which "evolved under the pressures of more complex social life until the species *Homo sapiens* appeared."

Many alternative plots have been imagined in the intervening years. But one idea crops up in virtually every hypothesis: The move from forest to open savanna was the first event in our story, both propellant and vector in the process of becoming human. Solve the key challenges of the open savanna, and the solution will, in time, unfurl all of the elements that define humanity. Natural selection will bind together these interdependent threads of humanness. Each will create the conditions for the other, reinforcing, tightening the fabric of our origin and existence. Open terrain—living on the savanna—thus created a precipice for the descent of all things human.

For Darwin and the mainstream of scientists after him, humanness is the interfingering of all the human traits in existence today, the integration represented by modern mankind—a whole defined by a certain union of characteristics. A creature less than this unity is somehow incomplete. It is recognizable as an ancestor because of its fractional humanity. We are beholden to the concept of humanness that we ourselves embody. It shouldn't surprise anyone that we imagine ourselves a culmination.

And so, whether Lucy was fully committed to terrestriality or whether the australopiths engaged in the art of tree-living is fundamentally a debate about our concept of humanness, how quickly the universals of human life were emplaced, and the strength and nature of the evolutionary tide that did it. Determining when the anatomical bond with terrestrial life was initiated is the lead sentence in the human epic. It is the real start of our origin, the spiral toward mankind.

Thus the central dilemma about the early australopiths: Were they mainly arboreal or terrestrial?

2

The environmental story of the late Miocene was written in the tense dialogue between directional change and oscillation. The oldest human lineage, I suggested earlier, disentangled itself from that of the African apes as the tides of habitat instability grew stronger. The oceanic forams indicate a leap in the pace and amount of environmental fluctuation, starting at least 6.4 million years ago, which then continued until about the 4.6-million-year mark. Mediterranean moisture pulsed to a rhythm that mirrored the orbital cycles of solar radiation felt all over the planet. At this time, close to the appearance of the first hominids, oscillations of global climate were larger on the average and more frequent than in any preceding era.

Oscillations in global ice and temperature appear to have happened less often between 4.6 and 3.5 million years ago. When they did occur, the largest oxygen-isotope shifts were up to 1.0 parts per mil, suggesting that broad reversals in climate continued.

The information we have shows that the earliest australopiths lived during sizable climatic oscillations on a global scale. But it is also true that these changes reflected, overall, an icier planet. During the coolest periods, temperatures dropped farther and farther. Earth was still in decline—colder, drier, and less forested, the conditions predicted by the savanna hypothesis. So the foram thermometer alone cannot help us decide whether the first tendencies toward bipedal behavior came from the drying and opening of habitats, or whether environmental variability played a more pivotal role in shaping this primal human characteristic.

We do know, however, that the basic architecture of australopith locomotion was no short-lived experiment. The curious alloy of the thighbone's long neck, inward-angled knees, short legs and long arms, the outward splay of the pelvic blades, and hints of a climbing forelimb served the hominids for a long time. With subtle variations in each species, this basic architecture is seen in Lucy's kind, *Australopithecus afarensis;* a later species, *Australopithecus africanus;* and even some early representatives of the genus *Homo.* The australopith skeleton was a fundamental overhaul of the basic arboreal design of early Miocene apes. In this sense, it appears to have been a long-lived, stable structure significantly oriented toward walking upright on the ground.

The questions that arise from this have long vexed anthropology: What were the environmental signals that propelled the origin and survival of the early australopiths? What were the challenges of their habitat? What benefits did they accrue from their new and persistent claim on bipedal posture?

The ocean-dwelling forams cannot help us here. Although reliable as a generic guide to environmental change, they lived too far away from the sub-Saharan terrain of our first ancestors. The harshly sculpted places that liberate fragments of early australopiths are our only true access to the mystery of these forebears.

A person breathes; a person swallows; one necessity precludes the other. You wake up and busy yourself according to personal ardors and agitations; you recline, slumber, slip across the threshold of an unconscious universe. On the face of it, we devote ourselves to basic processes of living that conflict yet mesh in cycles, according to the affairs and transactions of the antagonists. Life takes on definition by an involuntary merger of potential contradictions. In a given period, there is little choice but to search and to rest, to ingest and to fast, to mate and to seek solitude. But the poles cannot be occupied at the same time. One thing gives way to its foil; experience arises from reconciling the contrasts.

Such dissonances are spawned far and wide, well beyond the human course. They are essences in the act of nature. Midnight's darkness is forgotten in noon's intensity. Winter's barren calm opposes summer's fragrant zeal. Fertile rains arouse us because of prior drought. One reverses what the other provides. And all living things respond to both the predictable and the unexpected antipodes of nature.

While people tend to express these extremes as opposites, this is not an accurate portrayal of how the natural world proceeds. The dissonances do not deny one another. Organisms do not die in the inherent conflicts of their settings; they accommodate. They transform the contradictions into a range of variation, a degree of change that can be tolerated. This is central to the business of life on Earth.

The science of human origin has fabricated its own opposites. One such grand division is that between the terrestrial and the arboreal. The former is aligned with all things human; we are ground-livers. The first challenge we experience is a parent's extended arms, the summons to stand and walk. The opposing side takes residence in the crook of a tree, emblematic of an animal we no longer claim to be. Human versus animal is one of the fundamental distinctions in Western thought. Thus the states of anatomy and behavior that gain entrance to the human fold are held to be in conflict with those that define the animal, the apes from which we came.

I wonder, though, how the concurrent influence of both trees and ground can be denied a place in mankind's ancestry. In the fossilized milieu that surrounds the remains of australopiths we find evidence of both. The early australopiths held the patent that united these two opposite realms into one.

The early Pliocene site of Aramis offers an intriguing counterpoint to the savanna hypothesis. Fossil seeds and abundant tree-living monkeys found with the remains of *Australopithecus ramidus* point to a forest or woodland setting for what is possibly the earliest known hominid. Could it be, as some authorities have said, that early hominids thrived primarily in the forest rather than in open terrain?

Later in time, the Laetoli site offers a rather different setting—obvious signs of an open savanna governed by cycles of aridity. Soils rich in calcium carbonate and minerals formed in an alkaline environment mirrored a climate that was at least seasonally dry, incapable of supporting a continuous forest. Erupted ash occasionally blanketed the landscape. Largely unhindered by tall vegetation, winds blew volcanic sands over the region. Studies of the fossil pollen, antelopes, and rodents of Laetoli point to the dry savanna of the present Serengeti as an appropriate model of the ancient habitat.

Not all of the clues confirm this portrait, however. Giraffes and other tree-browsing herbivores were also abundant during the 300,000-year span when grasses, grazing mammals, and arid-adapted rodents pervaded the region. A dozen species of terrestrial molluscs imply a richness of trees, possibly a dense woodland next to open terrain. Expanding on the ecological reconstruction of Laetoli, Peter Andrews of the Natural History Museum, London, has drawn attention to this other body of evidence, showing that swaths of trees were at least as widespread as grassy plains. His studies confirm similarities with the modern Serengeti, but primarily with its woodland segment. The diversity of browsing, grazing, and meat-eating animals, the proportions of terrestrial and tree-climbing creatures, all point to a complex patchwork of habitats. It is impossible to divide this 300,000-year span into finer intervals, so we cannot be sure whether the grass-woodland mosaic was stable over a very long time or ricocheted between sparsely and densely wooded as a result of climatic change.

The sites of Hadar, where Lucy lived out her brief life, and Makapansgat, where we first meet the species *Australopithecus africanus,* yield more substantial clues about the moods of Pliocene environments. At both locales, grassy, forested, wooded, and bushy habitats dominated at different times.

Around 3.3 million years ago, forest and bushland surrounded a broad lake within the Hadar region. Leaf-eating antelopes, pigs, and elephants were the main residents at first. But over the next 500,000 years, the lake receded and expanded. The proportions of open-country grazers and closed woodland species shifted, as they followed the whims of regional moisture. Time and again, mixtures of closed and open vegetation recast the landscape. Sometime after the 3-million-year mark, a grassy plain dominated the area, but within 200,000 years, the climate evidently reverted back to the bushy, forested conditions that prevailed at the outset.

Like Hadar, the environmental record of Makapansgat further weakens the idea of a consistent, stable savanna. Makapansgat, which overlaps the age of Hadar, is a limestone cave dissolved into the high veldt of South Africa. Studies of its fossil finds have inspired an amazing range of interpretations over the years. The first geologists to work there reconstructed a sere, virtually treeless savanna. Later geologists reversed this view, emphasizing the heavy rainfall and consistent mat of vegetation that must have occurred outside when sediments were being deposited inside the cave. Competing interpretations of the fossil rodents, arthropods, pigs, and antelopes covered the same range, dry grassland to dense woodland.

Most recently, R. J. Rayner, Ann Cadman, and their colleagues have documented wide fluctuations in Makapansgat's vegetation in the seven meters of cave sediments they studied. At certain points nearly all of the pollen preserved in the cave came from trees, usually forest species and bushland euphorbias. The oldest deposit that contains australopith fossils was from such a time. These intervals were interrupted by others when trees were scarce and species of grass and open shrubs dominated the pollen samples. In a 1989 paper, Cadman and Rayner conclude: "The data show clear evidence for dramatic and fundamental changes in the climate, and that wet forest conditions were found just prior to and during australopithecine times. . . . The pattern of events in southern Africa at the time of the appearance of *A. africanus* is one of fluctuating climatic and vegetational conditions."

The most interesting thing, from our perspective, is that Rayner and his colleagues, in a paper published in 1993, drop the emphasis on fluctuations and declare instead that *africanus* was primarily adapted to the forest. Forest pollen certainly dominated at the time of this hominid's first visitation. However, *africanus* also shows up later in the Makapansgat sequence, where data on pollen, sediments, and fauna suggest a sparsely vegetated savanna, cooler and drier than the previous era. So it appears that *africanus*—like Lucy and earlier kin—experienced a diverse succession of environmental conditions over time.

Aramis, Laetoli, Hadar, and Makapansgat are four pins stuck in the distribution map of the oldest australopiths, but even on the basis of this small sample, it is untenable that these australopiths were *adapted to* the forests. I suspect that the "forest hypothesis" is as weak as the adaptation-to-grasslands doctrine that has dominated for so long.

Discoveries of fossilized fruit, pollen, wood, and snails in the Omo-Turkana region add to this picture. Fossilized fruits of the rain forest genus *Antrocaryon,* restricted today to central Africa, have been unearthed in deposits 3.4 million years old along the Omo River in Ethiopia, and samples of fossil pollen have been collected from about this same time. Together, the two lines of evidence indicate that a complex, oscillating mosaic of woodlands and grasslands stretched over the region, interrupted by galleries of river forest that swelled and contracted over time.

More than seventy species of fossilized wood from the vast Omo valley have been found, ranging in age between 4 and 1 million years. From one stratigraphic level to another the species vary tremendously with almost no overlap. To paleobotanist Raymonde Bonnefille, this means that floral change in the gallery forests was especially rapid and dramatic.

Fossilized snail species, known to be exclusive to African rain forests, have been gathered from rocks 3.4 million years old along the western shore of Lake Turkana. According to paleontologist Peter Williamson, their appearance coincident with forest fruits in the Omo, one hundred miles to the north, means that a wave of forest swept across the region around this time. Bonnefille disagrees. She argues that ribbons of forest linked East Africa with true rain forests to the west; the forest ribbons were prominent but localized elements of the eastern landscape. Like the grasslands and woodlands, they coalesced and then separated, time after time.

What we know about the fossil residences of early australopiths poorly supports the idea of a stable or ever-widening savanna. The context of early *Australopithecus* points to a more interesting environment—a series of reversals in any overall trend. The earliest lineages of hominids had to cope with these changes. I believe that this is what imparted the creative charge to human evolution.

The early australopiths found habitats that were less arborous than those of earlier times, but there is little to support the idea that the earliest bipeds were forced or pressured by any single environment, or by its sporadic pulse, in one consistent direction. These beings were not governed by confinement to open savanna or to forest. As much as anthropologists yearn to paint the setting of our oldest direct ancestor, it was the plurality of scenes that moved the drama of the australopiths along.

In this heterogeneity of settings, there is a novel solution to the debates about Lucy's bones, the quandary of australopith posture. The appropriate question is not adaptation to either grasslands or trees, but of response to variability itself, with trees and open terrain as elements in the varied topography of hominid emergence and survival. Versatility was the hallmark of this amazing risk of bipedal behavior.

Locomotion is about mobility. It involves shifting an animal's location and body position relative to something else. Independent of evidence from australopith sites, it is appealing to think of locomotion as the first line of reaction to a vacillating environment. The human lineage was initiated by a change in mobility, an adjustment to a kaleidoscopic habitat.

The four oldest ventures in bipedal behavior that we know about—*Australopithecus ramidus, A. anamensis,* and *A. afarensis* in the east and *A. africanus* to the south—each persisted for a span of geologic time. Their existence was measured in hundreds of thousands of years. That they could tolerate and adjust to new configurations of their habitat, to the tides of forests and grasslands around them, was pivotal to their endurance.

In accommodating to both trees and ground, any special bond an animal may have with terrestrial movement adds the advantage of flexibility, because trees imply the availability of solid ground underneath. The converse is not always true; the presence of ground does not assure the presence of trees. A total commitment to two-legged striding means rejecting the advantages of trees—food and protection, for example. Alternatively, strong commitment to old arboreal ways would be an impediment when distances between the groves widened.

During wooded times, even if the best resources were obtained by the most agile arborealists, terrestrial activity provided the broadest opportunities in the long run—access to both trees and ground. Even if natural selection gave arborealists a measurable advantage, the division of vegetation into a forest-wood-grass patchwork required passage through the uncomfortable terrain between swathes of trees. Movement on the ground prevailed—with bipedal striding as one method—if only because it allowed the maximum opportunity to practice an arboreal life.

It is true that all apes have some peculiarity about the way they walk on the ground. Chimps and gorillas, surprisingly, travel with the weight of the front part of their bodies on the middle bones of their fingers. Orangutans bear their swaggering weight on fists set ungracefully on the ground. Bipedal positions occur in all of these apes under certain life conditions. There are also the gibbons, who delicately tightrope on two legs, hands raised high in the air. But don't Lucy's pelvis and her other designs on two-footed gait require some *special* explanation?

The origin of bipedal architecture in the australopiths, cannot, I believe, be divorced from their entire, well-cemented blend of terrestrial and arboreal tendencies. Andrew Hill of Yale University suggests that bipedal behavior began in the repertoire of a tree-feeding ape, one who went to the ground to traverse "what initially at least were unrewarding and useless tracts of grassland." This proposal sets the savanna paradigm on its head. The australopith body plan, with all of its apparent conversions to bipedal posture, was oriented toward versatility of movement, committed neither to the ground nor to the trees exclusively.

Any account of bipedal origins is biased by the fact that today a two-footed gait appears solely in a fully committed terrestrialite, *Homo sapiens.* In our anxious pursuit of the real beginning of humanity, it has been hard not to read ourselves into the past. We tenaciously seek the imprint of our present ways in the anatomical ruins of ancestors. What we so willingly discern to be a skeletal pledge toward terrestrial two-legged activity may instead have been a long-term, persistent, and successful pact with versatility.

The bipedal orientations of the australopith skeleton were splendid accommodations, permitting flexibility and generality of movement. The early australopiths possessed a more versatile locomotor repertoire than the proconsuls, and were structurally better able to accommodate to both ground and trees than were the middle Miocene apes. And they were far less attached to a single style of locomotion than human beings are today.

The australopith body was an answer to a habitat that offered no guarantees as to the density of trees or the breadth of open terrain. It was an unconventional combination of adjustments to the swaths of grass that had to be traversed. These are the signals of humanlike striding we can now measure and observe in Lucy and her fossil brethren. The same adjustments were balanced by access to an arboreal wealth, the benefit of things in the trees.

The australopith body was thus liberated from an apelike architecture and its necessary attachment to trees—a true revolution in locomotion. Simultaneously, it was unindentured to the ground, not bound to the exclusive terrestrial gait that the human lineage, large in body and brain, was later to adopt.

These new-styled apes had a diverse repertoire of movement. In the bowl-shaped modification of the pelvis and the length of the toes, every australopith conveyed a history—an era of parceling and intermittent change in the architecture of the landscape on which they lived. They walked using two legs, and they climbed with four.

The australopiths possessed this distinctive blend of movement by at least 4 million years ago. Its initial development cannot have been influenced by later manifestations of the environment—at Hadar, Makapansgat, or anywhere else. Traits do not evolve with a prophetic eye on future conditions. Australopith versatility was spawned during an earlier era. Evidently it took shape in the presence of alterations equal to, if not exceeding, the range of habitats in which australopiths were disposed to survive later on.

This recalls a distant mirror, the ocean forams and the spasms of the Mediterranean crisis. The broadest environmental range was expressed between 6.4 and 4.6 million years ago—a long interval of vacillation in which, I suspect, the human venture gained its independence. If this view is correct, a final implication about the oldest human experiment assumes its first, rough form. Some parameter of nature was engaged besides the orthodox advantage ascribed to fecund apes in an ever-drying savanna. Something more than the progressive match of a two-footed protohuman to a certain habitat was at work. It was a response to the curve of environmental variation, integral to our ecological genesis, a calculus yet to be fully embraced in understanding our place in an evolved world.

3

In the great environmental decline of the Cenozoic, recurrent suites of distantly related animals emerged in the Americas, Eurasia, Africa, and Australia. Crescent-toothed browsers, big herd-living grazers, pouncing saber-toothed hunters, fleet social carnivores, cryptic rodentlike burrowers, and small stealthy predators evolved as independent experiments on different landmasses. Most regions also witnessed the rise of large scavengers—soaring birds, the long-distance connoisseurs of decaying meat, and powerful carnivores who snapped, cracked, and gorged upon the last morsels of the dead.

Diverse lineages in distant corners of the world display their own separate evolutionary dramas, often accompanied by an acute and surprising mimicry of one another. These parallels of evolution deserve consideration. The best way to reflect on the *uniqueness* of the human condition is to examine its opposite, the far-flung commonalities or reiterations of evolved history.

In every case of this sort of parallelism, there is a strange and surprising link between organisms of very distant kinship. Any parallel we observe between independent lineages inevitably manifests signs of a unique heritage. Because Africa is not Asia, and a marsupial carnivore is not a placental lion, distant parallels possess special peculiarities and paces of development. The hyenas of Africa and the bone-crushing carnivores of South America evolved from two quite different stocks—placental and marsupial mammals—and so possessed differences in reproductive physiology, metabolism, and anatomy, despite similarities in feeding behavior and related structure.

No one could mistake the wildebeests of Africa for the kangaroos of Australia. Yet both have done superbly as the large herd-living grazer on their respective continents. The long-necked, long-legged camel of Miocene North America, *Aepycamelus,* browsed on the canopies of trees and evolved simultaneously with the early giraffes of Africa. But *Aepycamelus* is a camel, not a giraffe by any stretch of biology. And the late Miocene prairie of North America produced fast, predatory meat-eaters impossible to confuse with the famed lion, cheetah, and wild dog of the African Pleistocene. In each example, the discrepancies are, in part, the result of a period of separation and distinct heritage.

The horselike *Chalicotherium* was a large, curious beast that evolved in Europe during the Miocene. It browsed off high foliage by rearing on its hind legs and reaching with its clawed forepaw to draw leaves into its gaping mouth. Its semi-erect stance and modified pelvis have parallels among the ground sloths of South America and the gorillas of Africa; similarly specialized, large, clawed browsers arose in North America (the genus *Moropus*) and Australia (animals known as palorchestids). Yet because each aspect of parallel evolution occurred independently, the chalicothere, the giant sloth, and similar beasts inhabiting different lands do not necessarily look alike. Their true similarity is in their ecological role, a convergence of conditions and opportunities shaped by global trends.

This ecological parallelism must not be mistaken for identity. Much has been written by Stephen J. Gould and others about the critical importance of *history* in evolution: Because each species is distinct due to the circumscribed reproductive behavior of its members, each lineage carries its own specific history, different from all the rest. Because entire groups of organisms are permanently segregated from other groups—the main branching points in the tree of life—the lineages in a case of parallel evolution can never really be deemed identical. Each carries the stamp of independent history. Disparities arise because separate lineages may diverge significantly before similar environmental trends begin to assert their gravitational pull. Even when environmental shifts are global, as in the Cenozoic, each region of the planet experiences change in its own particular way. Modifications are not necessarily synchronized across all regions. Their specific expressions are the result of the meshing of distinctive species, ecologies, and physical attributes of the landscape in each domain. They reflect the idiosyncratic deals between genetic mutation and all the events that enter into life—mating, death, and natural selection—which cannot be expected to write the same story everywhere.

A land bridge connecting two previously separate areas illustrates how prior history may affect evolutionary results. The first meeting between two distinct sets of species is a geologic moment of drama, especially if the organisms on both sides have already evolved similar ecological roles independently. New arrivals may be sufficiently different to fit in and raise the number of species in one or both regions. But if the similarities are too close, contact may result in the demise of many of the old inhabitants. These outcomes consistently testify to the distinctive character of each region and the species evolved in it. Parallels do not override separate histories. Indeed, the separate paths of evolving organisms make the parallels—the fact that they *do* occur—that much more curious and notable.

A comparison between the large grazing mammals of Australia and Africa illustrates another key point. Animals on separate continents may evolve a surprising range of similarities, which cut across many different biological groups. At the end of the Miocene, the biological group of kangaroos and wallabies, known as the macropodines, began to diversify into a host of grazing and browsing lineages. From time to time, the lineages multiplied and then were pared down by extinction. The macropodines today include twelve genera and fifty species, and at least eight extinct genera are known. There is also a wide assortment of parallels to independently evolved animals of Africa:

> By elongation of their fourth toe, kangaroos are capable of efficient, fast movement over wide terrain. Many groups of African antelopes and zebras also evolved longer toes and longer limbs than their respective predecessors, providing mechanical advantage in fast long-distance movement.

In the macropods, cheek teeth migrate forward as they erupt and become worn, being replaced by the new grinding surfaces of later-erupted molars. This odd characteristic, exaggerated in the very diverse kangaroo genus *Macropus,* is also found in the diverse Plio-Pleistocene elephants of Africa, in the modern elephant, and in certain lineages of pigs in which abrasive, grazing diets have prevailed.

The forepaws of kangaroos and their cousins are manipulative devices used in handling food, grooming, and fighting. The parallel here is not with antelopes or elephants or pigs, but with the higher primates, especially terrestrial monkeys and apes.

Over the past 5 million years, the consummate grazers of the African savanna—the gnus and hartebeests—split into twenty-two species, which are classified into nine living and extinct genera. By comparison, the single genus *Macropus,* which comprises the grazing kangaroos and wallabies, also originated about 5 million years ago and has fourteen living species and at least nine extinct species known from the past 100,000 years alone. Entire radiations, not just changes in anatomical structure, offer testimony to parallelism.

The kangaroos display ways of feeding, handling, and moving that emulate the survival tactics of distant lineages in foreign lands. They are usually thought to be equivalent, in the ecological sense, to the grazing ungulates of other continents, and indeed they show the same combination of large body size, wide mobility, reliance on grasslands, large groups, and rich diversity observed in the ungulates of the African plains. The kangaroos, however, manifest a broader array of affinities and peculiar and specific parallels with elephants, pigs, monkeys, and apes. In most cases, the similarities are independent responses to strikingly similar environmental pasts.

From these examples, we can discern at least three kinds of parallelism. The first is anatomical; it occurs when two distant lineages evolve very similar bodily forms, a process called convergence. The second is ecological; it arises when the same ecological role is fulfilled in two quite different ways—hopping and running, for instance, as methods of rapid transport. The third concerns overall evolutionary history, when distant groups of animals exhibit similar patterns of origin, extinction, and species diversity.

Parallel histories are an analytical scalpel we can use to dissect the influence of external context on the evolutionary paths of organisms. This can expose and check one lineage's response over time and place against that of other organisms. Each path of evolution is a kind of trial. By comparing the diverse probings represented by different species, and the outcomes of their existence, we can determine which trials worked and which didn't in the weathering temper of our planet.

Such comparisons provide a way to decode the vast number of distinct lineages that make up the history of life. Each branching path of evolution draws together subtle and surprising parallels with other, distant lineages that have experienced similar environmental regimens. A great deal in each unique path of evolutionary origin may be understood by reference to more general phenomena—first by finding similar developments in multiple groups of organisms, and second by studying the common external conditions in which these organisms evolved.

Comparing environmental histories worldwide, we would be lucky to find many parallelisms. And we might be amazed to discover large-scale trends in adaptation and lengthy stretches of climatic variance that apply across multiple landmasses and separate lineages. This is the fortune we discover in the archive of the late Cenozoic. It allows us to deduce the general factors that affected the existence, change, and death of species over that particular interval in Earth's history, and it helps us grasp the multitude of seemingly distinct directions of evolutionary change.

The spread of the kangaroo's hop over Australia's broad terrain tells of a more general problem—the opening of habitats at the expense of forests, due to drier and more varied settings. And each herbivore lineage that evolved higher-crowned teeth testifies in its own way to the decline from the rich, warm climes of the early Eocene. Recognizing such parallels of form, role, and overall history is a critical challenge in the explanation of evolution.

If parallelisms are a way of understanding the individual paths of species, what might we learn if we could apply this same method to the human lineage in all its vainglorious uniqueness, sovereignty, and separation from nature? What examples from history might penetrate the notions of singularity and superiority we passionately reserve for human beings? To what broad signals of environmental change can we ascribe both the responses of animals *and* the evolution of human uprightness, technology, culture, intelligence, and command of resources?

As analysts of our own deeply rooted childhood, we must invite such probings. The answers to these questions will be the hallmarks of our collective identity: a declaration of humankind; a statement about why the human lineage carries itself in the natural world the way it does. In these answers we will edge toward our origin as a phenomenon of nature, for ours was a birth with consequences and outcomes, like that of all other living beings.

4

It has always been tantalizing to think of bipedality, probably the oldest human invention, as the Great Leap that instigated an inevitable evolutionary process. It is tempting to think of two-legged walking as the dawn of the human ecological odyssey, the rich seed of human dominion.

Are there developments in the animal kingdom that parallel our bipedality? Humans are the only primates to move habitually on two legs, and the only mammal to walk with a smooth, repeated stride from one foot to the other. But several other groups of animals adopt a two-legged, largely upright posture as they move around, and it is useful to discover the advantages conferred by this locomotor approach.

Birds, for example, habitually stand or perch on their hind legs and devote their arms to remarkable transitions between land and air. Flying is their locomotor advantage. In kangaroos, the benefits of bipedal posture are conveyed in the speed and efficiency of two-legged hopping, without any apparent new function implied for the arms and hands. Gibbons and siamangs, known as the lesser apes, also adopt an upright posture in moving through the trees; in this case, the arms are used to swing from branch to branch, again, a locomotor purpose. These same apes are equally famed for their acrobatic two-legged walking on tree boughs; as they engage in this behavior, their arms and hands serve no function except balancing. Gelada monkeys shuffle in an upright position when moving a short distance between one clump of food and another. In doing so, they avoid the small but cumulative energy cost of switching between a bipedal crouch and their typical four-legged walk. On this basis, Richard Wrangham of Harvard University has suggested that upright locomotion in the earliest hominids also concerned efficiency in moving from one food source to another.

From these examples, it is apparent that upright or bipedal locomotion need not imply anything but a locomotor advantage. This way of looking at the first human invention runs counter to tradition. It has long been thought that upright locomotion originated because of some other benefit—the manufacture of tools, as Darwin believed; the ability to see predators in tall grass, as Louis Leakey once suggested; or as a way to avoid overheating in a hot, open habitat, as Peter Wheeler has recently urged. Each of these suggestions echoes the savanna hypothesis.

There is a more obvious explanation, where the benefits of walking erect are found in mobility itself. The distinctive two-leggedness of the early australopiths, I maintain, afforded a certain flexibility of movement. Its success resided in the opportunities it gave to adjust locomotor style to changing landscapes, an accommodation hardly ever seen in the living species of hominid. In the period between 6 and 3 million years ago, we have discovered highly variable habitats. Mosaics of forests, woodlands, and open plains were altered over time, varying from place to place. The advantages of australopith locomotion mirrored the mutability of these landscapes. Although examples of bipedal parallels among other animals are pitifully few, each case demonstrates that upright locomotion has provided an advantage in propelling a creature from one place to another. The key benefit and cause of uprightness does not need to be found in some other category of behavior.

If this way of looking at early hominid mobility is true, it breaks the imperative that has bound anthropology for many decades. Walking on two legs no longer signifies a fateful event causing the inevitable cascade of all things human.

Bipedal locomotion obviously had important ramifications. At one time or another it affected the way our predecessors fed themselves, their angle of vision, their avoidance of predators, the freeing of their hands, and their escape from the full onslaught of an equatorial sun. But these were secondary benefits, none of which robbed the early australopiths of the primal benefit of their style of locomotion. In the variable environs of the Pliocene, the first hominids were dedicated to a type of mobility that was tolerant of terrains both filled with and depleted of trees. Their peculiar upright style could accommodate the myriad landscapes that tested their survival over time.

So the departure of hominids from other animals was not an especially portentous one. The idea that bipedality was the beginning of anything particularly noteworthy in the history of life is a matter of pure hindsight. The peculiar trademarks of human dominion were all developed much later in time, removed from the settings and forces that parlayed an ape into a versatile biped.

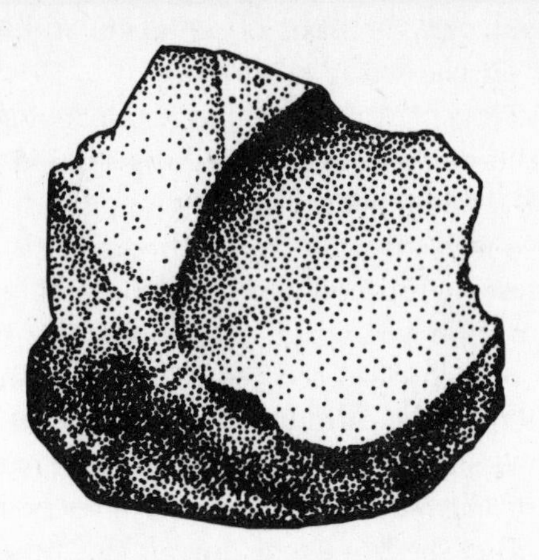

Certain fundamental activities are responsible for the impact humans have on their surroundings. Carrying and handling objects are crucial to our effect on habitats. Humans garner life's necessities by using hands and machines to transform objects. Simple implements made by carefully breaking one rock against another have long been considered powerful symbols of this behavior—the beginning of human technology. But it is now well known that chimpanzees, too, alter pieces of their environment to use as tools—termiting twigs, leaf sponges, and so on. Although other apes are less inclined to do so, their skill in making nests and handling objects suggests a general ape capacity to develop the use of tools in natural settings. However, as Mary Leakey pointed out in her pioneering study of the oldest stone technology, human beings are unique among primates in regularly *using tools to make other tools,* a breakthrough that is obvious today in the levers and chains and factories by which we disturb our surroundings.

Intimately linked with toolmaking is our tendency to transport things, another hallmark of human impact on environments. Not only do we make things that enable us to produce useful energy; we transfer the products and the technologies from place to place. Our societies have built fantastic webs in which the raw materials of one place and the energy resources of a second are brought to the industrial facilities of a third. Carrying things is the underpinning of this cardinal element of human life.

A direct result of human dependence on tools and transport is trash. We create leftovers and clutter the landscape with our slags and collected wastes. Herein lies the basis of pollution and the hazardous burdens we have laid on the world's habitats.

The earliest practices of stone knapping, transport, and trash making followed the advent of two-legged walking. Based on the oldest archeological records, these behaviors first appeared between 3 and 2 million years ago—well after australopith bipedalism had become established. The oldest definite evidence of stones chipped on purpose is a little over 2.5 million years old. At around the same time, monumental sways and branchings occurred in the human family tree.

Lucy's kind was no longer around. Her lineage, *A. afarensis,* and the older *A. ramidus* had succumbed, passing on their ways to new species. The southern *Australopithecus africanus* was already established in the African Transvaal. But the big event during this broad interval was the division among the australopiths. Their blood, their gene pools, bifurcated into the two great arteries of bipedal possibilities: the later, or robust, australopiths, and the larger-brained varieties that were seedlings of the genus *Homo.* Indeed, this dynamic era witnessed a minor radiation in our own limb of the primate family tree.

Contemplating the several thousand hominid fossils known from the time between 3 and 1.5 million years ago, and the numerous sites where they have been found, we can imagine ourselves in this extraordinary era and pretend to travel unlimited by time and place. We can rise high enough to discern an awesome scar in the African plateau. The terrain divides awkwardly along crooked lines where the Earth's hot plasma breaks the surface and tells a sweeping tale; the landscape has been shaken by unseen forces, regions have been consumed by flows of dark, inexorable burning and by ash.

In dreams of days and weeks, we find ourselves and observe an extraordinary bounty of creatures. As the vegetation and terrain produce their rich mixtures, a great diversity of beasts, large and small, are born into this land. Among them we catch glimpses of two-legged creatures, something like humans. We see them first one at a time, later in small groups. These bipedal beings are not all the same; they have different sizes and markings, and different responses to our presence. None of them seems completely at ease in open grass; neither are they attached to the woodlands.

These bipeds are the most human creatures on Earth. While what we see inspires a sense of awe, our hearts betray the lack of any clear or immediate feeling of identity with them. Which one of these beings would we call human? Which of these several kinds would engender in us the feeling of a special link?

We move swiftly along lengthy edges of grassland and trees. Each zone of the mosaic emits its special aroma. The plants and animals vary from one grassy or woodland patch to another. Neither grassland nor woodland nor the fringe in between is identical in any two places.

As we traverse this period, beginning 3 million years ago, we find that the savanna has no meaningful average. The grassland widens dramatically at certain intervals, but forests and woodlands still find refuge along the vast network of streams and rivers and ponds. These are the progeny of a landscape continually re-forming itself. Over time, the strings and patches of local flora are rearranged, made over.

Many of the animal populations split and fuse, coping with a dynamic geography and its rhythms. Groups of bipeds enter and exit by no apparent design, settling in or moving on according to complex signs that expand our present, limited template of nature.

The bipeds are pretty rare, so we begin to notice other lines of curious mammals woven into the environment's rich macramé. The large fauna consists of antelopes, pigs, elephants, horses, monkeys, carnivores, rhinos, hippopotamuses. In time we begin to note changes in the bodies of these animals, striking repetitions in the way they plan their growth and survival for another generation.

Herds of impala and waterbucklike antelopes browse the boundaries of grassland and woods. Groups of gazelles dart between large patches of grass, where they seem to feed, and denser thickets, where they seem to disappear. Ultra-large wildebeests give a strangely familiar chorus of grunts and hunker down in tall grass. By 2.5 million years ago, the grazing bovids radiate out over the landscape, mirroring the spread of grasses and herbs that result from crescendos in aridity.

The main disparity with modern savannas is the huge diversity of large mammals, a richness that mirrors the beautiful interweavings of forest, woodland, and grassland, which are revised in fitful rhythms. Pigs, elephants, zebras, giraffes, large monkeys, and carnivores divide into lineages of far greater number than are found in the present. The arbored zones are sufficiently varied to support at least four distinct kinds of browsing giraffes. Arboreal monkeys and saber-toothed carnivores take advantage of this woodland diversity.

The variety of open-country forms is even more peculiar. As we skip from 3 to 1.5 million years ago, at least seven different pig species can be seen in the open landscape as opposed to only two or three in the closed woodlands. Two main lineages of elephants thrive and give rise to new forms. The modern genus of African elephant, *Loxodonta,* is present until around 2 million years ago, then becomes rare. *Elephas recki,* related to the Asian elephant, is far more prominent. Its highly variable lineage dominates the large-bodied fauna between about 3 and 1 million years ago. At least four different kinds of large ground monkey inhabit East Africa during the Pliocene and early Pleistocene. The one-toed zebras and three-toed hipparions divide into their own diverse spectrum of horses.

Within these multiple lines, we find beautiful illustrations of a second kind of parallelism. Over time, many separate lineages evolve exaggerated structures for chewing plants. In many herbivores, the overall size of the cheek teeth increases, and the expanded molars form two large planes, upper and lower, on each side of the mouth. Large quantities of coarse foliage are minced efficiently between these two broad surfaces.

In numerous lineages, the crowns of the cheek teeth become taller, which prevents exhausting the supply of chewing surface too early in life. At least two lineages of pigs, the hipparion horses, and the grazing elephant *Elephas recki* exhibit this kind of change. Elephants increase the number of rough parallel ridges on their teeth; the *Theropithecus* monkey develops a more complex enamel surface; and *Equus* and other large mammals grow high, broad, complex teeth.

All this broadening, heightening, and increased complexity of the molars produces a massive increase in the structures devoted to processing food. The faces and braincases of these mammals must now support larger food-processing muscles and absorb forces generated by heavy chewing. The architecture of the skull becomes more buttressed and built up. A greater amount of bodily resources and growth energy is directed toward the biomechanics of eating. This suggests a stronger commitment to finding tough, hard, abrasive food, and lots of it. For certain animals this means a diet of fibrous grasses; for others, tough fruits or hard seeds; for still others, underground tubers and roots.

The human family experiences its own version of these developments. A variety of bipeds can be glimpsed. Hominid diversity after *A. afarensis* parallels that of other large mammals. The hominid radiation of species begets seven or eight distinct forms. Populations divide and spawn off. Some pools of hominids vary in intriguing ways from the anatomical icons that later define the main fossil predecessors of human beings.

In the southern part of Africa, there is a diversity of rather small species. Up until about 2.5 million years ago, *A. africanus* continued to thrive, but later gave way to a more massive-faced species, *Australopithecus robustus,* sometimes called *Paranthropus robustus.* In one area where *africanus* lived, we get a glimpse of a larger hominid with broad molars and a wide eye and nasal region, like *robustus,* but with large, protruding incisors. It is unclear whether this strange amalgam represents a different species.

Farther north, we see a short, bulky biped, squatting and eating something in the grass. He shuffles along using an outstretched arm. Tucked in his bent fingers is a stick blunted by wear. At one point, near some shrubs, he takes the stick, digs for about two minutes, and then levers up a sizable tuber, which he chews for a long time, spitting out what is distasteful. He sees us, rises, gives an amazed hoot, then utters a loud cry over his shoulder. On two legs, he runs with surprising speed toward a group of trees where he turns, calls again, and plucks a bunch of small green pods from a branch and shoves them into his cavernous gape.

During these excursions, we always sense something nearby. At times it could be something dangerous, mostly it is just movement in the grass, trees, and shrubs. On one such occasion, we hear a rhythmic rapping and spot a small biped hitting one stone against another. It is a female of a very short, muscular species. Her arms are long, and she is on solid ground near a series of tall reeds; a dense grove of trees lies a short distance beyond. A number of stones surround her, but she ignores them. Her focus is on an odorous antelope leg, which she has brought to this place.

We get a slight whiff of rotting flesh and notice numerous bits of bone littering the ground near her. She takes a rock and eagerly smashes the bare end of a joint. She's after the marrow and licks gingerly. Some moments later, she picks up a sharp rocky sliver she has made, slices along the rest of the bone, peels away sinew, cuts, and pulls a stringy tendon across her teeth.

Immediately, two other hominids with curiously short legs arrive, one carrying a rock. The other sees us. Without a sound, all three flee to the grove of trees, a smooth, tempered departure, as though this were commonplace.

We've had other glimpses of hominids this same size, perhaps this same species. They seem to tackle trees as they climb and disappear. Farther to the north is a much larger biped who visits pebble bars along dry stream channels and makes stone tools. This hominid seems the rarest of all. The face is massive, proportioned by a braincase larger than any we've seen before.

While we walk along, we find part of a skull, ridged at the top, broadly constructed below, from which the eyes face forward. Its huge, flat-worn plane from canine to last molar belongs unmistakably to a robust australopith. But matching hominids with their skulls is confusing.

It dawns on us that the stone-carrying biped used rocks as a kind of out-of-body food processor. The toolmakers could knap a cutting incisor, or whittle a stick into something like a canine tooth, or use a rock as a newfangled molar for crushing and grinding. They could make these new kinds of teeth, the first dentures, with their own hands. Early stone tools are damaged in ways that make it abundantly clear that cracking, crushing, and cutting were their main functions. The makers of stone tools discovered how to create such supplementary teeth outside the confines of their bodies.

The hominid whose skull we hold in our hand seems to have devoted himself to a different cause. His life's energy was directed toward growing a fantastic apparatus for processing food *inside* his mouth. The growth of this powerful facial and dental complex pervaded his life history. It signified an approach to growing up, eating, and living that seemed incompatible with that of the hominid who made new sets of lithic teeth. Probably this skull belonged to one of those long-armed bipeds who used the stick to lever up a tuber and delivered fleshy pods to its large jaws.

As we wake from our imaginary journey, let us consider how scientists actually interpret the array of fossil evidence that led to our dream. The skull we held in our hands belonged to the human known as *Australopithecus boisei,* one of the longest-lived of all known species of hominids. First discovered by Mary Leakey in the Olduvai Gorge, *A. boisei* belonged to a group known as the robust australopiths.

The term "robust" refers to the great size of their cheek teeth and the inflated appearance of their faces, buttressed against powerful chewing forces. The little certain knowledge we have of their bodies suggests that like the earlier australopiths, they were fairly small, perhaps from three feet seven inches to four feet six, and sixty-five to ninety pounds. Their brains were slightly larger than those of a like-sized chimpanzee, which means only about one-third the size of our own.

The oldest known robust skull, about 2.5 million years, is an extraordinary piece of anatomy. Discovered by Alan Walker on the west side of the Turkana basin, it is known as the "Black Skull" because of its dark fossilized hue. It has an enormous crest for chewing muscles near the back of the braincase, where later robusts have it farther forward. The Black Skull has the pillared, buttressed face typical of the robust clan, but it protrudes around the nose and mouth, unlike the flat, tucked-in physiognomies of the later robusts. For these reasons, many anatomists have favored distinguishing it, and several related finds from Omo and Turkana, with its own species name, *Australopithecus aethiopicus.*

It seems likely that the long, successful *boisei* lineage was a transformed descendant of *aethiopicus.* The strong, flattened face, the ponderous brow, the crested peak, the thick muscular ropes that moved the jaw, the massive molars, the molarized bicuspids, the thick bed of mandibular bone containing the teeth, all comprised the human parallel to megadonty, the sets of large, flat-topped teeth many other Pliocene animals evolved. The chewing factory the robust australopiths evolved suggests that they ate coarse, fibrous plants, and studies of the microscopic scratches on their teeth confirm the importance of these items in their diet.

As *boisei* thrived in the rift valley, a similar species evolved in southern Africa, *Australopithecus robustus.* It varies from place to place, and the species name *Australopithecus* (or *Paranthropus*) *crassidens* may apply to one of these variants. These southern australopiths represent a third and possibly a fourth species exemplifying the extreme chewing adaptations of megadonty. The southern and eastern varieties are easily distinguished, and evidence is mounting that the robust species from these two regions attained their powerful dental apparatus somewhat independently. If so, this reflects parallelism *within* even this small radiation of hominid species. *Boisei* and *robustus* represented the eastern and southern terminations of their proud lineages. The last known fossils of the robusts are a little more than 1 million years old.

The small, short-legged human, known from fossils discovered in the Olduvai Gorge, lacked the facial peculiarities of the late australopiths.

A fragmentary skeleton representing a female enlightens us about this species. The markings on her bones suggest a muscular body, though she was strikingly short in stature. Even a million years later than Lucy, the Olduvai skeleton known as "Lucy's Child" was only marginally larger. Her braincase was, however, somewhat expanded over that of earlier forms. With this combination of brain and body size, she was among the early bearers of the distinctive emblem of the genus *Homo*—an enlarged brain in relation to the body. This species is dubbed *Homo habilis,* and I have portrayed them as makers of stone implements who dabbled in the fine art of extracting fat and protein from dead animals. Minted in their skeletons were signs of the old arboreal ways. There is a strong possibility, suggested by their rather apelike body proportions, that they climbed trees.

The idea that a separate, larger species of *Homo* also existed around this same time has gained strong support in recent years from independent studies by anatomists Bernard Wood and Philip Rightmire. According to Wood, it should be named *Homo rudolfensis,* after Lake Rudolf, the old name of Lake Turkana. This species had the largest brain of all humans who lived prior to 1.8 million years ago. Long and stocky limb bones are known from the late Pliocene; if they belonged to this hominid, he must have attained a large body size, near the average for modern people. Moreover, as Wood has reported, this version of *Homo* flirted with the concept of large molar teeth, and his face was wide and quite australopithlike. This was an anatomical synthesis unto itself, a distinct lineage affected by both the dietary parallelisms of the time and the brain-oriented probings of the genus *Homo.* At its outset, even our own special lineage was susceptible to multiple trials and contemporaneous experiments in being human.

One other species, conceived late in the same period, overlapped with one or possibly both versions of early *Homo.* This was *Homo erectus,* the first hominid that we are certain possessed modern body proportions, and the first true migrant, who, by wandering beyond the African continent, announced the human aversion to confinement and presaged the global domain of his descendant much later on.

What can we make of all these species? The first point is obvious: The human line is no longer a line. The fossil record has given us something we didn't expect—not a simple route of progress, but a diversity of experiments. The second point is that the hominids were not alone in this respect. As the diversity of other mammals increased, hominids joined in on the act.

5

Having established the dynamism in the animal communities of 3 to 1.5 million years ago, let us probe the roots of technology, transport, and trash—the human triumvirate born with such disturbing innocence in that distant era.

Technology. The oldest human technology we know of is captured in stones of lava and quartz and is as simple as one could imagine. Hitting one rock on the edge of another makes a sliver of stone fly off. Its acute edge is sufficiently sharp to cut with. The larger piece of rock, or core, has a scar whose end is also useful for cutting and provides a platform for knocking off more sharp flakes. As this basic action is repeated, a core begins to exhibit multiple scars, as if a pattern were being made.

The hammerstone eventually shows signs of battering. Other rocks also reveal indentations and crushing. In the oldest known technology, Oldowan toolmaking, rocks were made sharp by percussion flaking, or were dulled and shattered by direct use. Superb collections of this oldest lithic technology have been studied for over thirty years, and the plethora of observations boils down to the question, What did stone toolmaking signify to the hominids who first tried it?

We've already seen one ramification: Stone toolmakers could start the process of digestion—chopping, slicing, and grinding—without being limited to the mouth's dental machinery. Compared to wooden implements peeled and broken by hand, tools could open even the toughest containers of food. And as the range of accessible foods grew larger, the same chipped rocks opened new channels to the diverse mosaic of savanna resources. Tooth-mimicking was the centerpiece of the first stone technology.

Transport. When hominids began to make tools from stone, the necessary raw materials could be found only in certain limited locations—at confined outcrops or in widely scattered clumps of stream cobbles. This meant that if tools were to be of use in any other part of the landscape, rocks had to be carried.

Rocks are heavy and have no caloric or nutritional value, so the time and energy spent finding suitable stones added a sizeable burden to the process of getting food. The new teeth were made a regular part of the food search only by bringing the stones and the foods together. Oldowan toolmakers solved this problem by moving resources from one place to another. The accumulation of stone tools on the landscape greatly extended the simple toolmaking practices known more widely among the primates.

Certain groups of chimpanzees face a similar problem today. In parts of West Africa, chimpanzees visit oil-palm trees to harvest edible nuts. In one study area, the chimps sometimes use locally available rocks as hammers and anvils to crack the tough outer casings. At certain trees they may require only a single stone in order to break nutshells against an exposed, anvillike root.

The chimps abandon these rocks at the bases of trees, revisiting and reusing them until nuts are ripe at a different tree. Based on studies by Christoph and Helwig Boesch, chimpanzees develop a mental map of where nut trees are located. They also remember where a stone hammer was last used. These places are visited, and if the hammer is still there, the returning chimp carries it to the next target tree, typically within one hundred meters. When fellow chimps have already taken the rocks to a different tree, the disappointed chimp continues to scour the places where nut-cracking stones might have been left, until one is found.

The chimpanzee nut-crackers' solution to the issue of getting the tool and the food together is to bring rocks directly to the food source. Nut trees do not change their location, so the hammers and the sources of food overlap precisely.

But what about the most ancient sites in the eastern part of the continent, where the only apes present were the bipedal variety? Only a few such sites, more than 2 million years old, have been excavated. The Gona sites of Hadar, Ethiopia, nearly 2.6 million years, are among the oldest known. Chipped stone tools have been found in and adjacent to the ancient streambeds. At the 2.4-million-year site of Lokalelei, West Turkana, early toolmakers used lava cobbles, some with very poor flaking qualities, retrieved from the streams that flowed near the site. It may be that as stone tools were first brought into the foraging process, they were used only in the immediate vicinity of the rock sources.

By about 2 million years ago, hominid toolmakers demonstrated a different solution to the problem of transport, seen at a series of well-studied sites excavated by Mary Leakey in the Olduvai Gorge. Various kinds of chipped implements and raw material, carried from places as far as one to six miles away, were discovered at the sites. Accumulations of animal bones, including fractured limbs and other body parts taken from animals that had died some distance away, were also found there. Cut marks made by stone tools, demonstrated in 1981 by Pat Shipman,

Henry Bunn, and myself, linked the two types of transported remains. Over a period of time, the toolmakers had broken and deposited the bones in the places where rocks had been carried. The oldest sites that give evidence of extensive stone transport, numerous parts of animals, and definite cut marks on bones are about 1.9 million years old.

The transport of acquired items was a turning point in the behavior of early toolmakers. At least two kinds of resources were involved—stone tools and foods that needed to be processed. Both were taken from their sources on the landscape and brought to common ground. The connection had been made at Olduvai, and probably earlier. Once the link was established, any movable item could be taken to a place where suitable rocks could be found. A stone flake, a core, and a hammerstone were surely carried around. But the advantage of tools was having them available whenever required. The solution was to drop lava cobbles and slabs of quartzite in various parts of the foraging range. It merely required the toolmakers to recall the places where these rocks had been temporarily discarded.

Chimpanzee nut-crackers suggest how this system got started. A stone or two was deposited at a favored tree. Perhaps a few other tools and rocks were left at a carcass, near water, or next to another favored resource. Over time, toolmakers were repeatedly attracted to these places. If food that required the use of tools was found, it meant a visit to a remembered location. Wherever stones were dropped, chances were raised that hominids would return. More visits meant more stones from distant places, the basic recipe for making an archeological site. This is one scenario that fits nicely with the archeological evidence from Olduvai. But however it was actually carried out, the dual transport of resources was a key marker in our ecological genesis that amplified the simple act of making tools out of rock.

Trash. By acquiring and carrying things to certain spots, early humans created zones of waste and clutters of objects both useful and useless. Besides giving jobs to archeologists, this behavior marked the beginning of a trait peculiar to human beings, which Glynn Isaac, who stimulated tremendous interest in the archeology of human origin, once referred to as the start of the human proclivity for making garbage.

The places visited again and again by Oldowan toolmakers were the first waste disposal sites, and they have stayed around for a couple of million years—which may give us pause about our artifacts and current pilings of refuse. Hitting rocks together and making clusters of debris

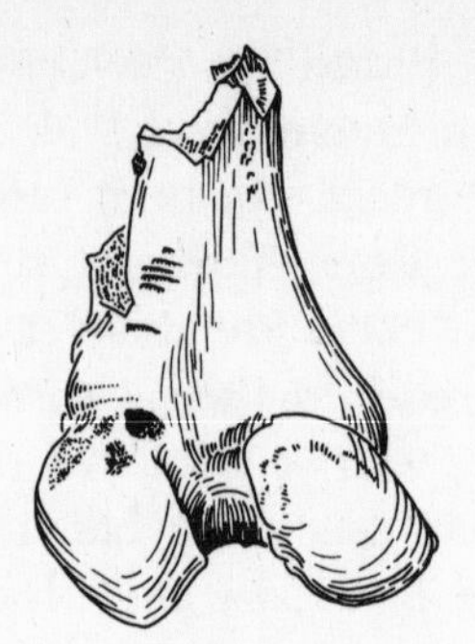

formed the tiny apex of the enormous taproot that nourishes our deep propensity to amass wastes.

Any innovation has its costs. What was it about toolmaking that financed its inherent costs? What advantages urged certain groups of hominids to seek heavy rocks and cart them from place to place?

If these primordial roots of human energy use were a Pandora's box, the box had to be cracked, sliced, and chopped in order to open it. And simple things emanated out of this box—seeds from a hard fruit, marrow from a large bone, pith from a tough stem. Simple stones wielded in a simple manner allowed a wider range of food sources than ever before.

Startling new forms of technology obtain a foothold in modern society when they reap new gains in energy or assist in securing some vital resource. These patterns of advantage may have been established by the oldest stone flaking. But we cannot yet compute the energy advantages of ancient toolmaking or be sure of the new sources of food that could be exploited. Because animal bones are readily preserved as fossils, and because of entrenched biases in our thinking, anthropologists often presume that meat was the first dietary breakthrough. According to this view, animals provided a new source of nutrition, underwriting the earliest toolmaking attempts. At present, however, there is no compelling reason to overlook the broad range of plant foods that could also have been reached by flashing those external lithic teeth.

Archeologists Nicholas Toth and Lawrence Keeley have searched for microscopic polish on the edges of stone tools. The only pieces to show microwear were small, sharp-edged flakes. The original materials on which these flakes were used included wood, soft plants (stems or grass), and meat. Even though their study focused on implements around 1.6 million years old—about a million years after the advent of stone flaking—Toth and Keeley have shown that access to meat was not necessarily the only motive for chipping a sharp stone.

Oldowan toolmakers eventually availed themselves of a strange new option—the nutrition locked up in large animals. Literally thousands of bone fragments, representing dozens of individual animals, have been unearthed in even small excavations. The sizes of the animals these represent are considerably larger than the occasional monkey or tiny antelope eaten by a chimpanzee or a baboon. While all higher primates eat an eclectic diet, none except human beings regularly seeks out animals the size of impalas, gnus, or even larger creatures. By the 2-million-year mark, human toolmakers had raided a totally new source of energy and nutrition, once the exclusive domain of large carnivores.

It is now widely accepted that this breakthrough was conducted in part by scavenging animals already dead. According to studies by archeologist Rob Blumenschine, marrow and meat can readily be scavenged in modern habitats of East Africa at certain times of year. While nothing rules out the idea that hominids occasionally dispatched an animal by hand or modest projectile, systematic hunting of large animals, clearly signaled at much later sites, is not apparent in the archeological record as far back as Oldowan toolmaking.

We can only imagine the first dashes into the realm of large predators and scavengers. Perhaps those initial events—stolen opportunities for innards and beast tartare—were spread out over thousands of years, an individual or a group here and there braving a hyena, a lion, or a sabertooth.

Spotting an unattended carcass, a biped seizes the moment, small teeth ripping at flesh already opened the night before, perhaps by a sharp-toothed cat. Hunger is sated by the moist, chewy tissues of a creature larger than any she had dared consider before. In that instant, neither the size of the animal nor the future importance of this activity means a thing. It's just that this large, unmoving, smelly thing has impenetrable bones layered with nutrition. You see these things standing together in the distance; now here is one on the ground, and it offers something to eat.

In this situation, the advantage of having a sharp sliver of lava, capable of slicing thick hide, is pretty obvious. It takes dozens of such slivers to get much meat from a large animal, and to extract nutritious marrow by cracking the bones demands a far heavier stone. A carcass may offer more than food; hide and ligaments also have uses in other aspects of life. Once again, bringing carcass parts and tools together in the same place at the same time seems crucial. Death sites of animals are far less predictable than plant-food sources. So knowing where stone tools had previously been dropped becomes an important fact of life.

When tense conflict occurs between large carnivores, it is usually around the death sites of their prey. The tensions translate into a kind of sharp-toothed pecking order as hungry scavengers wait their turn. Even in the face of such orderly conflict, dangerous motivations seethe around any carcass encountered by more than one meat eater. With the exception of vultures and perhaps the cheetah, all such scavengers were potential predators of the smallish toolmakers.

A surprising finding occurred some years back as I inspected several thousand fossilized animal bones unearthed from the oldest prehistoric sites of Olduvai Gorge, approximately 1.8 million years old. The diversity of animals, concentration of remains, and mix of skeletal parts made it quite evident that hominids had accumulated these bones some distance away from the original places where the animals had died. Long ago, something—or someone—had transferred bones from many different carcasses to specific spots on the landscape. At some sites, cut marks and the abundance of tools implied that human toolmakers were responsible. At others, tools were largely absent, and the gnaw marks on the bones matched those engraved by bone-eating hyenas.

A remarkable pattern emerged as I perused the sites where stone tools and fossilized bones coincided. Wherever hominids were active, the carnivores were there too. Large and small carnivores left distinctive glyphs in the same bony accumulations where toolmakers registered their own telltale marks. Both were drawn to the attached meat and marrow within the bones. These places were not the original kill sites, so scavengers must have followed the scent to the very places where hominids left their vital new dentures. This led me and a growing number of other researchers to believe that these sites were simple food-processing areas, lacking the social complexity of a modern home base. The benefits of fire, the safety of the family, and the swirling interactions of social life centered at the home were yet to come.

Still, these early toolmakers held in their hands a seed of no small destiny. They took one kind of object and, by transforming it, could open new sources of energy for their use. On two legs they initiated the transport of resources, and with no eye on future elaborations, they left behind the oldest piles of decayed and discarded artifacts.

We have yet to explore the environments in which these new facets of the human enterprise evolved. Since any novel behavior depends on surrounding conditions, we may wonder about the habitats in which stone toolmaking arose, and why this odd experiment by early humans endured. Toolmaking, transport, and trash collecting were not conceived for the sake of our present dominion. They were born in relation to environments of that ancient time, environments of uncertainty, prone to change.

6

Evidence from the polar regions indicates that something very important happened to the global environment of this period. Antarctic ice cores reveal a major expansion in the southern ice cap around 2.4 to 2.5 million years ago. The first major ice rafts are also recorded in the Northern Hemisphere at this time. The foram thermometer, mirroring the oxygen isotope signal of global ice, shows a critical drop at about 2.5 million years, a brief time after a reversal in Earth's magnetic field, known as the Gauss/Matuyama boundary. The abrupt shift in the ice curve resulted from an increase in the heavier form of oxygen, ^{18}O, as massive quantities of the lighter ^{16}O were stolen from the oceans. This occurred at the same time that vast ice sheets began to grow. On a global scale, Earth experienced a stepwise cooling, a dramatic prelude to a future frigid age.

This distant warning reverberated over the continents. About 2.4 million years ago, an extensive series of windblown deposits, called loess, began to form over much of China. Loess deposits meant that winds were able to carry particles of sand and silt over open expanses of low-lying vegetation. Great depths of windswept dust appeared for the first time in central Europe and western Asia. In the Andean highlands of South America, studies by Henry Hooghiemstra of the University of Amsterdam record a formidable shift in fossilized pollen, signaling a grassland expansion at the expense of forest. The interval 2.4 to 2.5 million years ago is the marker for all of these events. The tandem of cooling and drying was felt far and wide.

In a very real sense, this worldwide episode was a single distillation of the kind of environmental change commonly thought to have spurred critical developments in human evolution. The shift to open savanna seems to have occurred in one major upheaval, so the savanna hypothesis may be salvaged after all. Perhaps the expansion of savanna around 2.5 million years ago was the crucial nudge that set the human lineage on its course. An influential hypothesis advanced by Elisabeth Vrba of Yale University takes its cue from this important event.

Vrba's *turnover-pulse hypothesis* has stimulated a tremendous amount of research, spanning the disciplines of geology, paleontology, climatology, and evolutionary biology. A turnover pulse refers to any synchronized set of species extinctions and origins in many groups of animals over a limited period of time. Such bursts of biological change, according to this hypothesis, are initiated by shifts in global climate. Climatic events not only alter the conditions of life, they cause habitats to become more fragmented. As a consequence, animal populations also become more divided, and new species emerge as separated populations respond locally to their changing habitats. Vrba's hypothesis is that the origins of new species and new adaptive capabilities are ultimately linked to a major directional shift in global climate.

Vrba states that the change in the environmental settings of early hominids around 2.5 million years ago was exactly the kind expected by her hypothesis. Worldwide cooling and drying were sharply mirrored in African settings. Arid and open habitats infected eastern and southern Africa. Woodlands and forests ebbed. The Cenozoic decline was writ large in one wrenching leap, and as the wave of grass moved over the African landscape, natural selection favored creatures who could thrive in it.

Africa was an epicenter of a changing globe, and a significant measure of it was the fossil antelopes. New species of the wildebeests and gazelles, the dominant grazers of the African plains, made their appearance in the fossil record betwen 2.5 and 2 million years ago. According to Vrba, the antelopes of eastern and southern Africa underwent a major episode of extinction and new species origins at this time.

Change was not confined to the antelopes. In the Omo basin of Ethiopia, about 2.4 million years ago, rodents from dry grasslands replaced others typical of wetter, forested environments. As already noted, several groups of large mammals evolved higher-crowned teeth and longer limbs, indicative of an open habitat in the late Pliocene. These changes are also said to correspond to an extension of grassland around 2.4 million years ago. Shifts in fossil pollen in the Omo-Turkana region and in the highlands of Ethiopia also hover around this date. The challenge of a widening savanna left its mark.

Vrba suggests that our own intimate phylogeny succumbed to this same directional change. As wooded habitats fragmented and dissolved into open vegetation, bipeds spawned new and independent examples of the hominid condition. Behaviors crucial to the human enterprise evolved with the spread of a cooler, drier habitat. Our own genus, *Homo,* was a founding member of the new savanna biota. Stone toolmaking and the dental machinery of the robust australopiths evolved as adaptations oriented to the resources of the drying, opening landscape, evolutionary events that centered around the global climatic change 2.4 to 2.5 million years ago.

The turnover-pulse idea proposes, first, that global temperature fell precipitously. Second, that this event caused the spread of arid grasslands within the savanna patchwork of Africa. And, third, that the growth of these grasslands prompted synchronized change in hominids and other animal populations. The regional division of populations led new species to arise, while the force of natural selection caused new adaptations to evolve. The hominids were converted to live in open terrain.

Vrba's proposal leans heavily on the savanna hypothesis: Directional change to an open habitat caused profound alterations in human origin. The idea of a turnover pulse is appealing for other reasons. It appears to account for the era's dynamism. Evolutionary events between 3 and 2 million years ago, according to Vrba, were focused on a single climatic alteration, and included all the events described earlier in this chapter—the formation of new species, parallelisms across many lineages, and vital developments in the ecological history of our own lineage. The concept of a turnover pulse during the late Pliocene has, moreover, received resounding support from climatologists, who have discovered in their local and global archives evidence of significant change about 2.5 million years ago, just after the Gauss/Matuyama magnetic boundary was imprinted in sediments around the world. The concentration of climatic and biotic change at this special point in time is the core of this hypothesis.

Vrba has done a remarkable job of scrutinizing her own hypothesis, but three critical questions test its relevance to human origin: What was the nature of this climatic episode 2.5 million years before the present? Were global change, local shifts in habitat, and developments in hominids precisely correlated? Does directional change provide a stronger explanation of hominid evolution during this period than environmental fluctuation? Let us put the turnover-pulse hypothesis and the thesis we have begun to develop here to these tests.

A curious pattern can be seen in our investigation—whenever Cenozoic environments became cooler or drier, they manifested greater instability at about the same time, or immediately before the main shift. This happened with the cooling event of 2.5 million years ago.

In a paper widely cited by advocates of the turnover-pulse idea, Nicholas Shackleton and his colleagues showed that large icebergs formed in the northern oceans for the first time about 2.4 million years ago. Because of recent redating of the Gauss/Matuyama magnetic reversal, the time of this change was probably closer to 2.5 million years ago. Before this date, the ice system of the planet was largely confined to Antarctica, and this single ice pole controlled the heat gradient between the equator and the poles. After 2.5 million years ago, however, the North Pole also developed large ice sheets and began to play a major role in governing Earth's climate.

Overshadowed by this dip in the oxygen isotope curve is a series of intense oscillations. According to Shackleton's paper, global climate varied a great deal before 2.5 million years ago, affecting the signal of ice fluctuation in the North Atlantic. Indeed, a wider range of fluctuation in the isotope curve began at least 300,000 years before the critical date. With the approach of the 2.5-million-year mark, the range of variation rose even more.

The curve below, from Shackleton and colleagues' original 1984 paper, shows that the dip at 2.4 to 2.5 million was not a permanent change, but part of a longer period of increased instability. On occasion, the isotope readings rebounded to levels that existed prior to the dip.

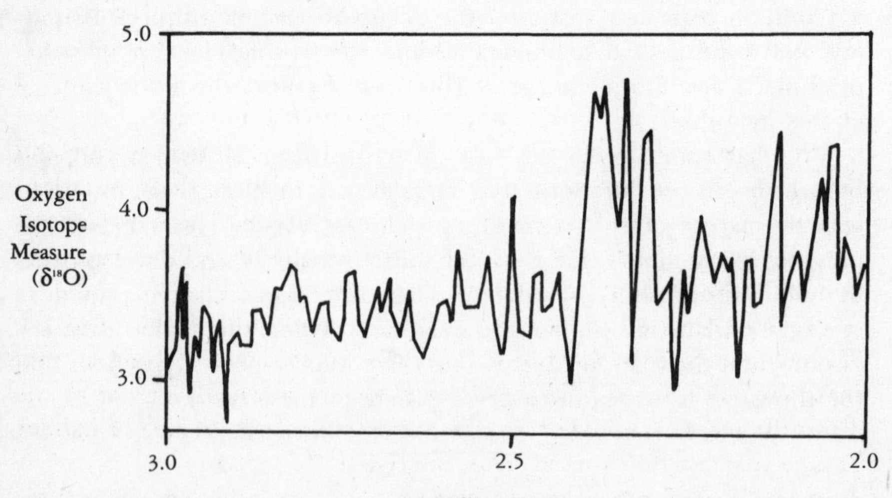

Isotope measures in other ocean cores portray a slightly different trend and pattern of oscillation. According to a Pacific Ocean core published in 1981, climatic variation began to widen as early as 3.1 million years ago. The widened pattern of oscillation occurred up to 2.1 million years ago, when even larger variations began to occur.

In 1984, geologists R. Stein and M. Sarnthein examined several other ocean cores drawn from the Atlantic Ocean, and inferred that deviations in the climatic curve increased in both frequency and intensity between 3.5 and 1.9 million years ago. They concluded that much of this fluctuation occurred 500,000 years on either side of the 2.5-million-year mark—that is, the entire span between 3 and 2 million years before the present.

The volume of terrestrial dust blown into the ocean is another indication of continental drying. Measuring sediments buried off the west coast of Africa, William Ruddiman and his colleagues found that the amount of terrestrial dust in the ocean increased around 2.5 million years ago, supporting the idea of increased African aridity at that time. However, many large-scale oscillations occurred in the quantity of dust that fell to the ocean floor. Ruddiman attributes the large, repeated shifts in the dust curve after 4 million years ago to "an increase in the amplitude of arid/humid cycles" over geologic time, with a major increase in fluctuation, including greater aridity, at about 2.5 million years ago.

Further clues come from a long terrestrial record often cited by the turnover-pulse proponents. Henry Hooghiemstra's record of fossil pollen in the high plains of Bogotá, South America, displays a significant drying at the 2.5-million-year Rubicon. Around the same time, a change also occurred in the oscillation of climate and vegetation. The pollen record indicates many small-scale variations in temperature and flora after 3.5 million years ago. Around 2.5 million, these variations became less frequent but of far greater magnitude. The key event of 2.5 million years ago was actually part of a lengthy spell of deepening fluctuation in global climate.

Besides climatic factors, volcanic and tectonic forces controlled the habitats occupied by hominids. There is no evidence to suggest that these irregular, more catastrophic events were unusually potent at the 2.5-million-year mark. Rather, they exaggerated the degree of habitat change that hominids faced time and again.

In the Omo-Turkana region of East Africa, volcanic eruptions periodically sent enormous clouds of ash and debris over the landscape, 4.2 and 1.4 million years ago. At least nine such events blanketed an area of more than ten thousand square miles. In some cases, erupted ash was carried eight hundred miles away to the Gulf of Aden. Even the smaller eruptions repeatedly altered the vegetation and soil chemistry over a wide area.

The Omo-Turkana study shows that arid savanna increased over time, but the water budget of this vast region was controlled mainly by tectonic activity, changing the flow of the Omo River and causing lakes to appear and disappear over spans of 100,000 years. The forest and woodland area surrounding these water sources was susceptible to significant, recurrent alteration.

The environments of eastern and southern Africa became drier overall, but the key global event, shifts in local habitat, and changes in hominids were not precisely synchronized. A pollen study of the Omo-Turkana region by Raymonde Bonnefille, for example, points to a shift to grassland habitats between 2.3 and 2.4 million years ago. In this same region, changes in antelopes, pigs, and other animals took place over a few hundred thousand years around the 2.3-million-year mark. Because of astronomical effects on Earth's climate, a difference of 200,000 years is a very important one, encompassing several reversions in the climatic trend.

The first appearance of robust australopiths, stone tools, and possibly the genus *Homo* precede the pollen shift in the Omo region. The oldest known find of a robust australopith is the Black Skull, *Australopithecus aethiopicus,* found in deposits several meters beneath a volcanic tuff dated 2.52 million years ago. Independent studies of the skull suggest that its peculiar combination of ancient and advanced traits must have branched from the hominid family tree independent of *A. africanus,* who was known in South Africa between 3 and 2.5 million years ago. Since *aethiopicus* has certain key similarities with *A. afarensis,* the former may have evolved from the latter. And since *afarensis* is unknown in the fossil record after about 2.8 million years ago, this date may provide a minimum age for the origin of *aethiopicus.* The robust lineage of East Africa arose before the abrupt savanna expansion posited by the turnover-pulse hypothesis.

The oldest stone tools known from the Omo-Turkana basin are dated about 2.4 million years. A few early archeological sites in this region and in Zaire and Malawi appear to be about 2.3 to 2.1 million years old. Yet the oldest definite stone implements, from Hadar, Ethiopia, are nearly 2.6 million years old. While these early implements correlate roughly with the proposed turnover pulse, they precede other signals of drying and grassland expansion in East Africa by at least 200,000 years.

New dates on a fossil fragment discovered in 1967 near Lake Baringo, Kenya, suggest that the oldest known member of the genus *Homo* may be 2.4 million years old. It has been argued that this appearance coincides approximately with the oldest stone tools and is related to the dramatic spread of open savanna. Other researchers have difficulty accepting the Baringo fragment's age and identity, and see definite signs of *Homo* no older than about 2 million years. But even if *Homo* and flaked tools first appeared around the 2.5-million-year date, they lie within the period of intensified oscillations that began prior to the abrupt cooling event. Was it habitat oscillation or a single megashift to cooler, drier habitat that figured more importantly in the turnover of species and adaptive innovation among the hominids?

Five distinct species of early human may have inhabited the Omo-Turkana region between 2.6 and 1.8 million years ago. The turnover-pulse idea has remarkable appeal partly because it sees a dynamism in the hominids similar to that in other lineages of mammals. According to Vrba's hypothesis, the spread of savanna caused a phenomenon known as "environmental forcing"—the division and extinction of lineages under pressures associated with a particular climatic trend.

It seems to me that vacillation—reversals in the trend—created a much more appropriate setting for this mutable era of human evolution. This, too, is a kind of "environmental forcing." But the critical factor was the increasing degree of habitat fluctuation, not a single directional change. Hominids and other organisms had to survive the spread of grassland beyond its previous limits; but our grasp of this vital period in human origin is not complete without also asking how these ancestors fared in subsequent moist and warm phases. The survival conditions of human forebears cannot be rendered by any one portrait, or any single type of biome. The change in climate at 2.5 million years ago may have been dramatic, but the longer interval of heightened instability between 2 and 3 million years ago also cast its influence on hominid evolution.

Robert Foley of Cambridge University has underlined the importance of habitat fluctuation in the turnover of late Pliocene species. In his scenario, cycles of change caused habitats to break up into smaller, more distant fragments. Populations became disjoined and isolated for varying lengths of time, adapting to different local conditions. Periodic dividing and coalescing of habitats seem more essential to a continental radiation of species than any unidirectional habitat trend. Environmental variability—the intensity and rate of fluctuation—may well have held sway over the persistence, splitting, changing, and extinction of lineages.

Two proponents of the savanna-pulse idea, George Denton and Michael Prentice, shed a different light on the timing of the climatic decline. According to their data, the main climatic event in the late Pliocene was 2.4 million years ago, during the buildup of ice at the North Pole. Again, this marker seems too late to have affected the earliest appearance of robust hominids and stone tools. Denton and Prentice suggest, however, that climatic deterioration began around 2.8 million years ago and culminated some 400,000 years later. Their broader "pulse" model is based on oxygen-isotope data and sea-level change. In recent papers, Vrba has embraced this revised time scale, but she still stresses the power of the linear trend, the rise of savanna, as the critical source of change in hominid evolution.

The environmental code of the era was a sequence of dots and dashes. Over this lengthy interval, numerous stops, starts, and reversals occurred in ice buildup, ocean depth, and planetary temperature. It was literally a tide with ebbs and flows. As water was taken up and released by polar ice, sea level fluctuated at least fifty and up to one hundred meters, constricting and reexposing huge areas of the continental shelf. The amount of moisture monsoon winds carried to the continental interior varied with distance from the ocean. Swings in sea level, which took place over tens of thousands of years, altered the atmospheric moisture blown into the continental basins occupied by hominids, affecting the rainfall and vegetation of these areas.

Prentice and Denton interpret the change in sea level between 2.8 and 2.4 million years ago "as primarily reflecting Antarctic Ice Sheet fluctuations with a minor but increasing component attributable to Arctic ice caps." It is reasonable, then, to suggest that fluctuations regulated the pattern of environmental change. Cooling and drying occurred as part of a widening spectrum of oscillation over at least several hundred thousand years.

How do toolmaking and other curious innovations in hominid behavior tie in with this picture of an unsteady environment? In trying to decide why early attempts at flaking and carrying stone persisted in the human repertoire, we are stuck with our old conflict. Were these behaviors adaptations to the open savanna? Or did they mainly provide certain hominids with a way of dealing with environmental variability? Let us consider what these novel behaviors were good for.

It has occurred to many of us who are curious about the oldest stone tools that the hominids who processed their food partially *outside* of their bodies were the most liberated of all bipeds from the demands of any single type of environment. The new dental opportunities made possible by stone tools meant that the toolmakers could transcend the status quo of any single habitat or slice of time.

In any particular environment, there is a recognizable pattern to how foods are distributed, and certain foods occur more abundantly than others. Therefore "optimal" foraging routes and "best" food sources can be defined in any given habitat. But even in the brevity of an annual season, the savanna offered a changing buffet. The ultimate test of the toolmaker way of life came as environmental extremes were felt over longer spans of time. I believe that lithic toolmaking persisted as a useful strategy precisely because it enabled a hominid to switch to different resources when the old ones were gone. By chipping rocks, certain hominids discovered a new form of versatility. A heavy stone and a sharp-edged flake meant that a tremendous variety of items could be opened, cut, or crushed. Changes in food supply were handled by making implements capable of processing whatever kinds of food happened to be available.

For all their simplicity, fractured rocks offered a kind of buffer against natural shifts in the resources affecting survival and reproduction. A chipped stone first became valuable when it performed some specific task—cutting a tough plant stem, sharpening a stick, or slicing an animal hide. I believe that stone flaking endured not because it encouraged this specific task, or because the original environment of toolmaking continued to influence the hominids, but because cutting edges and pounding stones allowed potential differences between an arid grassland and a moist woodland to be reconciled. Stone flaking afforded a resilient means of obtaining needed resources in the full range of environments.

Similarly, carrying food and stones to the same sites buffered potential conflicts between one environmental state and another. The first time lithic tools and animal bones were brought together took place in some specific environmental setting. But the act of transporting items persisted and developed because of the benefits it provided in a changing environment. Once there was transport, the fact that stones and a particular food resource changed from being fifty feet apart to being two miles apart did not prevent the toolmakers from bringing these two critical resources together. Transport, much like stone toolmaking itself, enabled hominids to survive unexpected changes in the distribution and abundance of natural resources.

Although the earliest Oldowan flaking is more than 2.5 million years old, the deeper implications of this primal technology may not have been discovered until 2.2 to 1.8 million years ago. Sites in this time range preserve the oldest evidence of wide stone transport and tool cut marks on animal bones. Deep-sea cores, drilled from the North Atlantic and the tropical Pacific, reveal two unusual times of oxygen-isotope fluctuation between 2.2 and 2 million years ago. Two ocean cores from the western Atlantic indicate an intense oscillation at about 2.2 million years ago, and another just before 2 million years ago. On a global scale, relatively wide climatic oscillations coincided with the advent of distant transport and dietary changes in human toolmakers.

On land, this period is less well known. The best environmental record we have from this era, from the Olduvai Gorge, is 1.8 to 1.7 million years old. At the outset of this interval, Olduvai was dominated by moist woodland vegetation and high lake stands, an environmental state that was maintained for about 40,000 years. Within the next 10,000 years, extremely arid conditions developed. Lake level dropped significantly; fossil animals indicate semi-desert conditions; the pollen record shows a shift from river forests to steppe. Shortly before 1.75 million years ago, lake levels were again on the rise, and closed vegetation and humid climate prevailed, until the trend was reversed by another abrupt shift to aridity, which piled up windblown dust about 1.7 million years ago.

The vegetation was drier and more open at the end of this sequence than it was at the outset, but the trend was not smooth. Aridity was interrupted by moist periods; cooling, by warming. In the shadowed passages of time preserved in the gorge, we see that toolmakers practiced their craft over the entire range of habitats. Long-term and gradual shifts in rainfall and temperature, occasional deluges of volcanic ash, tilts in the landscape from rare and awesome fracturing of Earth's crust, together must have caused impressive rearrangements in the resources of survival. Yet there is hardly a stratum in the most ancient depths of Olduvai Gorge where the imprint of stone toolmakers is absent.

To accept the anthropological dogma that the toolmakers flourished in drier, open habitats is to imply that human adaptation was molded primarily during periods of aridity. Alternatively, the environments of every time span were involved; natural selection produced relative success in all climates, wet and densely vegetated spans as well as periods of dry grassland. Knapping and stone gathering cut across the fluctuations. The hominids were geared toward surviving the entire environmental panoply.

Although it invited the attention of other meat eaters, the inclusion of large animals in the diet also succeeded as a buffer to change. Any major climatic shift or geologic event would have disturbed the vegetation, altering the abundance and location of plant foods. Herds of large animals would also have been affected, but animals do not manifest anything like the dramatic variation in nutritional quality and toxins associated with different kinds of plants. To eat an open plain's zebra is much like eating a woodland's kudu. To digest an underground tuber is not the same as eating the soft pulp of a fruit. Eating large animals helped stabilize the diet when climate and vegetation changed.

To select from the mosaic's cornucopia, it was necessary to follow the dilation and contraction of edible items. Some foods were given up and others adopted as the occasion arose. Bringing resources together required the toolmakers to maintain a good mental map of their milieu, and also the capacity to change the template. The toolmakers had to respond to the opportunities of moister terrains as much as they did to growing aridity.

The search for mobile resources, cunning links between rocks and food, carrying things over a distance, avoidance of predatory carnivores—comprised the package of the earliest stonesmiths. The combination was both dynamic and critical to the future of their descendants. Each of these novelties became especially significant in the next round of human origin—the rise of a large-bodied, large-brained, sweaty, long-distance nomad, who would later become the sole survivor, the stem from which all future humanity would emerge.

Before turning to this survivor, *Homo erectus,* let us briefly conclude our study of "savanna forcing."

> A worldwide decline in temperature and moisture caused the opening of savanna environments. Many organisms in the late Pliocene of Africa display dental adaptations to this trend. The prominent environmental pulse 2.5 million years ago was, however, actually part of an overall increase in habitat instability. Repeated environmental reversals may have played a more important role in the origin of *Homo* and of novel behaviors associated with this branch of hominids.
>
> Speciation, extinction, and shifts in African habitats did not necessarily correspond to one directional change in climate. Multiple, wide oscillations in habitat are a more plausible explanation for the dynamism of this period (multiple species origins in various large mammals) than any single event of directional change (savanna forcing).
>
> Toolmaking, transport, collecting resources and wastes, and dietary change make sense as ways by which certain hominids accommodated to changing environmental regimes, not merely as adaptations to dry, open savanna.

This way of looking at the environmental history of the late Pliocene casts a new perspective on the conditions of life faced by human ancestors, and on the manner in which natural selection acted upon the organisms of that critical era.

7

While humanity's lineage is said to have been pulled by a directional shift in environment, we find instead that the ancestral habitats of hominids fluctuated dramatically. Certain behaviors new to hominids—early instances of technology and other vital rudiments in our ecological temperament—first appeared in the midst of long-term oscillations in environment. These primordial developments were, in a sense, adaptations to variability, not responses to a single model habitat or the consequence of some narrow tunnel of adaptive pressure into which an ape entered and emerged a human being.

Spread of grasslands and fluctuation in habitats were the two motifs of the African Pliocene. Expansion of dry savanna was a critical part of the widening range of environmental oscillation. This is not to say that toolmaking and transport were adaptations to dry savanna per se, but as the Pliocene came to an end, the savanna had settled in for the duration, and the mark it left on the surviving species of hominids became deeper and deeper.

The next phase in the ecological genesis of humankind was transacted in the lineage of *Homo erectus.* At full height the early members of this species were nearly six feet tall. Working at a site called Nariokotome, west of Lake Turkana, Richard Leakey and Alan Walker's team excavated a beautifully preserved skeleton of early *H. erectus.* Initially discovered by their colleague Kamoya Kimeu, the skeleton is that of a male whose second molars had just erupted. In modern human terms, this makes the boy about twelve years old; he was five feet three inches tall when he died. The cause of death is unknown but occurred near a quiet bog inhabited by catfish and visited by hippopotamuses.

The Turkana boy tells us that early *H. erectus,* besides being a tall biped, had arms and legs proportioned like a modern human's. For his height, his arms were not long like those of Lucy, Lucy's Child, or so far as we know, any other prior hominid. He lacked the apish details that, in earlier bipeds, suggest occasional tree climbing. The legs and hipbones of *Homo erectus* were buttressed by tremendous thicknesses and bulges, which denote a body geared toward endurance walking and running. An exclusive pact had been made with the terrestrial realm, and the boy's legs were equipped to cover the ground in strides protracted in both length and hours.

Body size affects the ecology of all mammals, including human beings. Compared to small animals, large ones tend to be more mobile and to range over larger territories. Brain size also increases to maintain the original bodily and mental functions of the organism. From one species to another, body enlargement drags the brain along with it.

Here we begin to detect the extraordinary interweaving of the factors tied up in the evolution of human beings and other organisms. The growth of a larger brain or a bigger body requires longer periods of fetal growth. Higher primates—monkeys, apes, and humans—generally possess large brains for their body mass, and so have relatively prolonged periods of gestation and infant care. For a big increase in size—as strongly implied in the comparison between four-foot-tall *habilis* and six-foot-tall *erectus*—the period of infant nurturing was probably significantly extended.

Prolonging the early years of childhood affects the entire life cycle of an organism. Large bodies and brains and long periods of infant dependency are associated with longer life spans. Longer life spans cultivate the benefits of learned behavior; experience begins to loom larger in the maturing of the individual. A longer period of infant care intensifies the social bonds between individuals, and these, too, magnify the opportunities of learning.

Body size is thus the maypole around which the multicolored streamers of life history are wound. Maturation, longevity, brain size, learning, and social life are prominently entwined. The ties that complete the dance have to do with food and metabolic rate. Growing large obviously necessitates increasing the amount of food and, usually, the area of food search. And so we come back around the circle—bigger organisms have greater ranges of mobility.

Homo erectus's cardinal features were long legs, lengthy stride, and bones built to withstand the rigors of forceful, prolonged two-legged gait. These changes endowed this hominid with a degree of durable mobility unknown to his predecessors. A wider search for food was critical if a bigger body oriented toward muscular endurance and a larger brain were to be maintained. Appropriate strategies that facilitated the widening of *erectus*'s range had already been established. The dietary versatility sparked by Oldowan toolmaking was critical. The ability to make the link between far-flung stones, plants, animals, water, and other resources was already in place. *H. erectus*'s mobility simply enhanced the system of transport. A broader home range inaugurated a more extensive and intricate series of paths between resources. Coupled with the transfer of stones and food, greater mobility demanded a more extensive mental map of the terrain, which is probably one reason why brain size increased beyond what automatically went with a larger body.

As Leslie Aiello of University College, London, has elegantly shown, the metabolism needed to grow a larger brain had to be borrowed from some other part of the body. The heart, liver, and most other organs that consume a considerable amount of energy are too vital to borrow from. The intestines are another matter. Aiello's study shows that the gut of an herbivore is long and large, while that of an animal with a high-quality diet, including meat, is smaller and metabolically cheaper to maintain. Human intestines obey the latter pattern. Thus, eating meat may have had a dual effect. First, it increased the density of food resources, providing a rich new source of nutrition. Second, it may have enabled a certain amount of basal metabolic energy to be diverted from the gut for other purposes, including the growth of a larger brain.

In *Homo erectus* we see the origin and amplification of fundamental elements in the ecology and behavior of human beings. Not one step in our reasoning has gone beyond the natural principles of body enlargement, or the social and ecological implications when any large mammal species is compared with a smaller one. But the fact that all of these normal ramifications occurred in an omnivorous, toolmaking, resource-transporting, debris-collecting biped had tremendous consequences.

Out of this swirling interplay, one question begs to be answered: What stimulated the increase in the first place? One factor may have been the episodic spread of grasslands which would have multiplied the distances *H. erectus* had to travel between favored patches of food. Lengthening the legs and increasing the stature would have allowed this habitat change to be taken in stride.

There was also another factor in the ancestral path of *erectus*—something already in his background; something inherent in his use of tools, his occasional search for a carcass, and his transport of resources. It was a factor all animals have to deal with, yet none so strongly as a biped who has stuck his nose into the brashly odorous atmosphere of a dead antelope and has the temerity to introduce these scavengeable advertisements to the same places where he keeps his dentures. I am referring, of course, to the risk of attracting predators and becoming prey.

Animals have several options in dealing with predators: avoidance, trying to outrun them, and the formation of large groups, which raises the number of watchful eyes and the chances of defense. An alternative ploy is to evolve a larger body, which enhances defense against the large carnivores, eliminates the danger posed by smaller ones, and so reduces the overall risk of becoming prey.

We could argue incessantly about whether hominids formed larger groups or ran a little faster. The one thing we know is that by 2 million years ago certain hominid toolmakers were seeking large animal carcasses. And in some settings, such as Olduvai, hominids, cats, hyenas, and canids converged on the same places and left their telltale inscriptions on the same sets of bones.

Even if they did not actively hunt, the occasional search for carcasses meant that toolmakers entered the arena of the large predators. This placed a tremendous premium on growing larger, with all of its complex, cascading effects. Considering everything that body size touches, cause and effect inevitably entwine; they wrap around each other until, eventually, the maypole is hidden and its primal purpose is mysteriously cloaked in the measured rounds of evolution.

As a bipedal toolmaker took on the metabolic burdens of a large body and somewhat expanded brain, there was an unavoidable union of all the mandates and advantages this entailed: a larger foraging area; a longer growing period; the consequences of a longer life and more intense social bonds. In the length of his bones, the boy from West Turkana—or at least his next of kin—held a seed of enormous consequence: the anatomical means to be a traveler. This initiated one of the most important developments in the ecological saga of early humans. By 1.5 million years ago, *Homo erectus* experienced the journey beyond the distant hill and the location of his clan as some populations left the continent of the rift. They were the first humans to have done so in sizable numbers.

Claims have been made, often on slender evidence, that hominids appeared outside Africa prior to this time. In southern France, chipped rocks resembling the oldest tools are claimed to be 2 million years old, as are stone tools discovered in Pakistan. Artifacts from Yuanmou, China, may be 1.7 million years old, and a fossil jaw and tooth possibly as old as 1.9 million years was found recently in Longgupo Cave, China. Unearthed at Dmanisi in the Republic of Georgia, a fossil jaw similar to that of the Turkana boy may be 1.8 million years old. And ages of 1.8 to 1.6 million years have recently been linked with *Homo erectus* fossils found some decades ago on the island of Java.

Some of these claims are dubious, but further work may convince skeptics about others, particularly the Dmanisi jaw and the Java hominids. There is no reason to discard the idea that early human populations seeped out of Africa before 1.5 million years ago. Shortly after this date, there is more evidence of hominids in Asia. A site in Israel named 'Ubeidiya preserves stone tools and fossil animals 1.2 to 1.4 million years old. In China and Indonesia, hominids were clearly present by about 1.2 million years ago.

Whether we praise or lament the modern impulse to overstep the finiteness of our surroundings, we see in *Homo erectus* the spore of such recent exaggerations. As early as 1.5 million years ago, toolmakers left their mark in high altitude zones of Ethiopia previously uninhabited by hominids. As they moved through subtropical and temperate latitudes, nomadic groups must have encountered plants, animals, and landscapes new to striding bipeds. When hominid toolmakers arrived in eastern Asia, they inhabited areas as far north as the Nihewan Basin, about equal to Beijing in latitude. In Europe, the early record of tools and human fossils is far more subtle and poorly dated, perhaps not much older than 500,000 years. By that point, some hominid populations had severed the ancient bond with tropical climates. They faced a terrain and a temperature of which no descendant of an ape had prior experience.

Little is yet known about why hominid populations emigrated to new regions. In textbooks, the geography of human ancestry is usually relegated to a map with arrows directed out of Africa. According to paleontologist Alan Turner, one of the few people who have studied the matter, the early migrations of *Homo* into northern zones generally corresponded with the dispersal of lion, leopard, and spotted hyena, which he calls our "fellow travelers." Since these are large carnivores, meat and marrow may have provided the common advantage to both migratory hominids and their disconcerting companions. Dependence on animal protein and fat could have facilitated passing through lands of unfamiliar plants with their unknown toxins and indigestible compounds.

Unfortunately, we know nothing specific about the ecological opportunities that may have linked Africa and eastern Asia during the early Pleistocene. Researchers have yet to compare distant fossils and prehistoric sites to discover whether *Homo erectus* cut his own path to Asia or followed changes in habitat open to many other mammals. The essential point is, however, clear: the itinerant passages of *Homo erectus* foreshadowed the current state of the human species, the ultimate infiltrator and colonizer of all continents and terrestrial biomes on the planet.

The advantages of mobility, large body, and opportunities for travel, all exposed the long-distance migrant to potentially fatal hazards. Wider mobility ran the risk of longer separation from dwindling water sources. A larger body was harder to cool, and a larger brain greatly raised the metabolic demands on the entire body and the chances of overheating. Since humans deal with heat stress by the evaporative cooling of sweat, the solution to overheating merely worsened the problem of thirst and water supply.

It is not surprising, therefore, to see in the fossils of *Homo erectus* the oldest adjustments to such risks. Chris Ruff of Johns Hopkins University has shown that African *erectus,* for all his muscularity, maintained a relatively narrow pelvic width, and so possessed an elongated body consistent with life in the tropics. This suggests that early *erectus* met and dealt with problems of heat stress and water loss the same way that modern peoples of the tropics do.

Human skin is nourished by a vast network of blood vessels that transfer heat from the body's core to the surface, and has an enormous density of specialized sweat glands that cool the skin by evaporation. Evaporative cooling is helped by our skin's "nakedness"—actually a miniaturization of hair, not its loss. This unique combination effectively alleviates heat stress, preventing outright exhaustion, brain failure, and metabolic breakdown. Unfortunately, no skin or other soft tissue of any early hominid has been preserved to test our strong suspicions that *Homo erectus* was the first hominid to face daily the deadly hazards of overheating and water loss in a dry, open environment.

There is, however, one prominent clue in the bony subtleties of the skull. *Homo erectus* was the first hominid to possess a projecting bony nose. Earlier hominids retained the basic design of the ape nose, two openings in line with the overall contour of the face. In contrast, *erectus* had certain features of nasal projection that are found in all modern humans: The margins of the bony opening were raised, the nasal bones were widened, and the two were joined together in an angled nasal bridge. The upper half of *erectus*'s nose was expanded, as in all living human populations. The raised nose has important implications, detailed by Robert Franciscus and Erik Trinkaus of the University of New Mexico. It is a water-retention system, a vital little design prompted by the need to humidify dry, inhaled air.

The new nasal shape of *Homo erectus* expanded and made the moist, upper-air passages far more complicated. As in modern humans, the new nasal complex helped to moisturize dry air—crucial in the functioning of the lungs. At the same time, it created an intricate maze to catch and interrupt the hot, damp air expelled from the lungs. The water in each exhalation was condensed onto the mucus lining and retained to humidify the next breath. In modern human noses, this system has been amazingly efficient, significantly lowering water requirements. Its appearance in *Homo erectus* indicates that water retention was a critical factor in life, and offers a subtle clue that the hot savanna had laid its burdens on the human lineage.

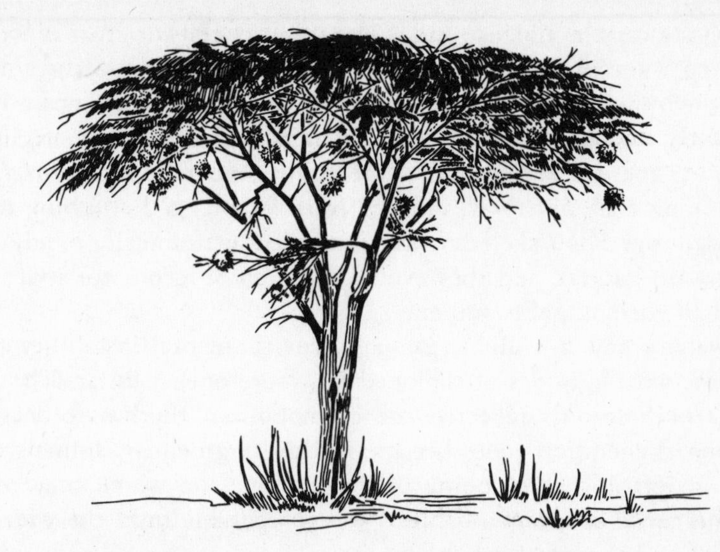

Based on anatomical appearance and his entire hot, sweaty, nasal portrait, *Homo erectus* was adapted to aridity, and indeed, every habitat indicator from the time of the Turkana boy indicates the entrenchment of dry savannas in sub-Saharan Africa. Environmental conditions worldwide seem to have settled into a more stable pattern. According to the oxygen isotope curves, global fluctuations in ice and temperature leveled off somewhat. After 1.7 million years ago, hominid habitats in the rift valley and the Transvaal stayed consistently open and arid.

This seems to imply that *H. erectus* was a product of an overall trend—a bipedal, large-brained toolmaker adapted to the savanna. At the same time we must acknowledge the defining measures on which the lineage of *erectus* was built—behaviors that endowed the genus *Homo* with a means to endure considerable inconsistency in the conditions of survival. These behaviors were the insurance policy with which this human ancestor entered the next era alone. The savanna hominid could hardly have guessed what was coming next, could not have foreseen the environmental extremities that would accrue in open terrain, or known about the rigors on continents beyond that of his origin.

8

The collectivity of human beings asserts a profound, torrential force. Our manipulation of things is sharp, cunning. Our alteration of the planet has a certain violence and seems to upset the structure of nature. In tracing the roots of this conflict, we have sought the settings under which the human condition was first fashioned. The oldest known hominids lived in a diversity of local settings and may have originated during a time of relatively extreme environmental oscillation. They moved in an adaptable way, ranging from four-legged climbing to bipedal walking. Their skeleton was rearranged anatomically to adjust to variations in habitat, and they evolved a union of locomotor styles that worked in their mutable milieu.

Between 3 and 2 million years ago, environmental instability again increased, with episodes of coolor, drier, more open habitat. The local settings of hominids reflected the cosmopolitan rhythms of nature's alteration. Vegetation zones became further fragmented; animals faced a finer geographic partitioning of the resources on which they relied. The emergence of savanna habitats was inseparable from the widening of spatial and temporal variability.

Animal populations divided, coalesced, and reorganized their means of survival. The diversity of anatomical forms greatly increased. As patches of open grassland spread, a multitude of mammals devoted to eating its resources evolved. Most clans of large herbivorous mammals gave rise to species having larger and more complex molar teeth. These were the mortars and pestles with which the coarser, tougher, more fibrous vegetation of the savanna would be treated. The robust australopiths transacted the business of eating plants in this same manner.

As some lineages reacted to the overall direction of change—cooler and drier—new behaviors emerged that succeeded in dealing with the variability of landscapes and resources. One or more species of hominids dabbled in the rocky outcrops and cobble-rich streams that dotted and traversed the landscape. By knocking stones together, certain hominids explored new ways of accessing the diverse range of resources in the savanna-woodland mosaic. Toolmakers transferred rocks from one place to another. Eventually this produced effective ways of gathering the nutrition that required tool processing. Transfer of food and stone to the same places meant that the heterogeneity of Pliocene habitats could be dealt with.

Stone flaking and transport, culinary involvement with a gazelle, discard of the refuse, all began as a modest correction in the uncharted course of hominids through time. Stone tools represent the oldest definitive modification of the environment by humans. Yet nothing seems to hint of any activity out of sync with nature. Perched in the present, we usually see the oldest altered stones as the beginning of a fulfillment, the small bang that made inevitable our massive, thrilling technological achievements of the present. But for hundreds of thousands of years—durations that dwarf the mere centuries in which our arsenals have been developed—the earliest stone toolmakers did little else to the raw materials than chip sharp edges and barely change their appearance by using them. I believe that we should not impose our own desiderata of improvement, competition, and progress upon this apparent "good thing" that the oldest stone flakers cared so little to mess with.

Two million years ago, certain lineages still retained some of the original fluidity of human locomotion. Some bipeds kept a small, long-armed body design, which facilitated alternative modes of movement. At some point, our own ancestral lineage surrendered this founding principle of hominid flexibility and signed on to a more specialized manner of bipedalism, linked with enlargements in body size. Locomotor versatility was sacrificed only after new approaches involving tools, transport, and accommodation to changing resources had evolved. These behaviors provided even broader insurance against the risks of an uncertain environment.

Homo erectus was the first real terrestrial specialist among the hominids and gained a kind of buoyancy by virtue of greater RAM in his portable skulltop computer. This hominid was the wanderer, unleashed from the basins of his origin. The untethering of African hominids was the preamble to the geographic story of human origin and the basis for the global scale of human influence.

In the dust of our *erectus* ancestors lie the fragments of other experiments in being human. The memoir of other hominids resides solely in cracked, hardened bones, and it is we alone who are left to write it. We must tell a tale that damns the other bipeds to extinction, a disharmonious story of a million years of survival and persistence that ended without issue. Our instincts and diligent search for meaning hide the possibility that this tale, the death of a hominid, has any significance for ourselves. Yet the extinction of the australopiths seems somehow connected with our own survival, with nature's ultimate choice of the genus *Homo* as the sole representative of hominids in Earth's biota.

Shortly after 1.2 million years ago, the robust australopiths were dead. This meant that at least three species of early human no longer existed. We don't know why they became extinct. On the basis of the sheer number of fossil finds, the robusts were prevalent between 2.5 and 1.5 million years ago. A census of that period might have registered more of these big-toothed, crested bipeds than any other species of hominid, including those with the incipient large brains.

At least one robust lineage experimented with an expanded brain. In East Africa the species *A. aethiopicus* was probably the forerunner of *A. boisei,* which became prominent 1.8 to 1.5 million years ago. Brain size seems to have expanded nearly 30 percent in about 700,000 years, within that line of kinship alone. We commonly assume that a trend toward larger brains guarantees human success. But this increase did not assure the robust australopiths' survival.

If the mark of success is longevity, these hominids were immensely successful. They were an integral part of the East African scene for more than 1.5 million years. The megatoothed *A. boisei* thrived for over half of that duration. In the south, the lineage of robust australopiths lived for more than half a million years. By comparison, modern *Homo sapiens* has been around for no more than 200,000 years, and many authorities would halve that estimate.

It's not at all clear whether the late australopiths should be called specialized. Their teeth suggest that they could chew just about anything by pulverizing it. They may have ingested foods other than plants, and even the herbivorous part of their diet may have offered a varied menu. One clue in understanding their survival, and possibly their demise, is that whatever the robusts happened to put in their mouths, the burden of heavy, muscular chewing followed. Eating anything required them to move their bulky jaws, activate their massive chewing machinery, and apply their broad planes of enamel—even to the most delicate berries or soft marrow. This was inherent in the approach evolved by the late australopiths, their signature and abiding commitment.

Homo erectus was on a completely different track with one foot in the ecological domain of the carnivores, and the other in the varied terrain and flora of an explorer. Two hominids could hardly have been more distinct in stature, anatomical proportions, pattern of growth (brains versus teeth), and geographic tendencies. For this reason, I agree with those who think that direct competition with *erectus* had little to do with the extinction of the robusts.

Other species, including the pig *Metridiochoerus compactus* and the monkey *Theropithecus oswaldi,* committed themselves to a large-toothed approach. But while the megadont humans met their demise, other savanna herbivores lived on into the mid-Pleistocene, evolving even larger molars and bodies than their predecessors. The massive chewing machinery of the late australopiths was emblematic of the savanna path enjoyed by these other mammals. Oddly, that particular path contained no guarantees of future survival for a biped.

At least two species of *Homo* lived during the boundary period known as the Plio-Pleistocene—probably even three, given that *H. habilis* and *H. rudolfensis* evidently persisted after the lineage of *Homo erectus* had evolved. If these bipeds truly bred independently, as distinct species do, only one of them could have been ancestral to the lone living hominid.

The question is, Why? Why the division into species and then extinction? We know about the extremely close genetic relationship between chimpanzees and ourselves. What is the meaning of extinction among even closer, bipedal kin? What allowed some hominids to endure while others died out? Some had large brains; all were bipedal. Some, if not all, had the capacity to use tools; some made tools of stone. Each lineage developed, culled, and compounded certain essential qualities of what it has meant to be human.

The intriguing, if hypothetical, question in my mind is: What if the rich Pliocene diversity, the parceling of humanness into a number of different species, had continued? What if the pruning of the family tree hadn't happened, and we were left with more than one survivor? There is an unmistakable sense that something important was going on, well over a million years ago. Extinctions were momentous turning points in the future of the hominids and of Earth's biota. The human family tree was shaped by the death of lineages as much as by the successive reproduction of human traits. Perhaps if we could understand the persistence of some of these experiments and the extinguishing of others, we would learn a difficult truth about the origin and conceivable future of our species.

What about our finely threaded line of descent, stemming from an early australopith, leading into an early version of *Homo,* then through the long saga of *H. erectus,* and on to other archaic versions of humanity that followed? We are a species now barely grown up and have left behind these progenitors of our long childhood. Our human parentage is extinct. While they gave rise to us, there is no denying that *Homo erectus* and the other species have vanished.

For some of us, there is fear in recognizing that we are derived from them, evolved. We ignore them without any conceivable sense of the bridge they have provided between us and the rest of nature. They have passed on and now lie dormant in the universal genealogy of our kind, which they engendered. All that remains of them are ripped and yellowed Polaroids in the farthest dusty corners of our house.

If we are to find our ultimate kinship, which is guaranteed to be with the animals, it is for us to discover it, to rebuild the sense of connection, and to discover in ourselves, beneath the thick presumptions of special destiny, a grasp of their existence. Not worship; rather a sensation, a feeling, the buzzing of natural process in our veins equal to that of the apes and other organisms around us.

What is the meaning of our forebearers' extinction? Are we not different from these predecessors? Won't our fate also differ? We must scrutinize such questions, turn them over and over and pass them on to others to judge. We register our opinions, but the real verdict rests in the hands of our future surroundings, much as it did for the robust and migrant versions of our kind.

Neither African australopith nor migratory *erectus* could possibly have foretold the emergence of our ecological dominion, our impingement and alteration, the spread beginning in pockets of domestication and intensifying as deep imprints on the soils of the planet and its waters and air. Unlike our predecessors, we pull levers and use propellants of our own making, moving the conditions in which we live according to principles of our own.

The history behind these imperatives has still to be addressed. Until now, we have witnessed only some of the seeds of the current state of mankind's affairs. We have yet to encounter the exact milieu in which the antecedents of the modern personality matured and changed into something quite different. We enter now this last era in our making, 1 million years long.

CHAPTER V

SURVIVAL OF THE GENERALIST

I

TODAY, people all over the world seize the rights to terrain and foreclose on nature's resources. We alter the extent and structure of habitats as no previous force on Earth has done, stripping the tropical rain forests to mere pockets, redirecting the rivers to serve our extravagant needs, exterminating species with startling swiftness.

Extinction, we are told, used to be rare, until mankind began to bruise the planet's ecosystems. It is generally assumed that our transformation of the environment has occurred with unprecedented speed. The change from the original to the current state of affairs began 10,000 years ago, with the planting of crops and controlled breeding of animals. Our disturbance of nature, now measured in mere hundreds of years, if not decades, is widely believed to be unnatural and unparalleled.

Virtually all environmental insults stemming from the human arsenal are considered dissonant to the relative stability of nature. The past was a system in balance. The environment had a kind of status quo against which conservationists compare present rates of extinction and degree of habitat destruction, the unrelenting turnover of nature to human use. According to this view, culture and nature are at odds, and it is humanity's overwhelming cultural manipulation of nature that now dominates.

Four elements of human dominion are central to the environmental dilemmas we now face: the technological facility with which human beings alter habitats; human aptitude for finding and using new environmental resources; human capacity for relatively rapid cultural change, seemingly discordant with nature; and the global effect of human activities.

Technology provides the leverage behind our alteration of the environment. Our use of tools, equipment, and structures is tied to the flexibility of the human brain and human cultural abilities. Tools and learned behavior, coupled with large brains, are the basis of human progress, providing access to all kinds of environments.

Life's social, linguistic, and neurological foundations allow the successes and failures of human environmental exploits to be registered with each generation. Our cultural abilities are a source of innovation, an apparently bottomless cup into which modern societies dip, often at the environment's expense. This facet of culture gives credence to the idea that it is the central munitions factory in our conflict with the balance and status quo of nature.

The global impact of humanity comes from the worldwide dispersal of human beings. Our mysterious, collective potential to migrate, to go beyond previous limits, obviously rises out of our technological capacities and the cultural products of our mental and social architecture.

All of these key elements—technology, intellect, culture, and worldwide impact—are intertwined, stitched into a kind of unity distanced from nature, and were nurtured in a prehistoric womb of human ascent until the water broke and the large-headed ways of altering the world descended on Earth's original environments.

By now, however, our investigation has gleaned enough to make us wonder about this vision of nature, and of ourselves. We must realize that this view of humankind—our conflict with nature, the speed of human alteration, the disjunction between human activities and the original state of the environment—rests on our view of nature, our grasp of what life was like before the last few thousand years of human dominion.

Is the prevailing portrait of the environment correct? Does our brutal disruption of the present vie with the status quo of the past? How much do we understand about the origin of technology, culture, and our current environmental dilemmas? We find ourselves facing the greatest mystery of all: How much do we understand about our own presence on Earth?

2

Let us return for the moment to Olorgesailie. Walking in that land of white dust, even the casual visitor notices a roughly chipped, symmetrical rock lying on the ground. Known as a handaxe, its crude, oval shape was imposed on stone by early people over an immense period of time.

For more than 1 million years, the handaxe was the dominant product of human technology. All the axes, which occur by the thousands at Olorgesailie, look rather similar. Distributed over Africa, Europe, and parts of Asia, the handaxe was the main stone design created and manufactured by hominids of the mid-Pleistocene. In our present world of rapid-fire technological advance, it is unthinkable that any single manufactured item could endure, much less remain dominant, for so long.

Monkeys and apes solve problems with insight and innovation. It is reasonable to believe that *Homo erectus* possessed an even higher degree of mental aplomb. He was certainly one of the handaxe makers, possibly the main hominid responsible for them. Yet the tools he left behind had little to do with novelty or active imagination. That the toolmakers made handaxes out of the best and the worst of rock types suggests a certain degree of adaptability and also a kind of uncompromising conformity, a rigid formula whose persistence is mind-boggling to us today.

The handaxe's hypnotic charm seemed to reside in its versatility. The million-year spell was reasoned, useful, and widely applicable. In one excavation at Olorgesailie, our team uncovered a skeleton of the ancient form of elephant known as *Elephas recki.* It was surrounded by hundreds of sharp lava flakes that, nearly 1 million years ago, had been used to butcher it. The flakes were boldly struck from handaxes, yet not a single intact handaxe was left behind. The handaxes were blade dispensers. The used blades were discarded, while the dispensers were carried away for further work.

At other sites, handaxes were the main implements left at the butchery sites, and experiments with modern ones indicate that they were good butchery tools. But their versatility is shown by microscopic studies of European flint tools less than 500,000 years old, indicating that handaxes and their flakes were also used in cutting wood and tanning hide.

The handaxe was a complex interweaving of versatility and monotony. Its 1-million-year term of service is, however, almost unthinkable by today's standards. We tamper with just about everything we create or lay our hands on. The handaxe people just kept on making handaxes.

Part of the canon in the study of human origin is to claim that the

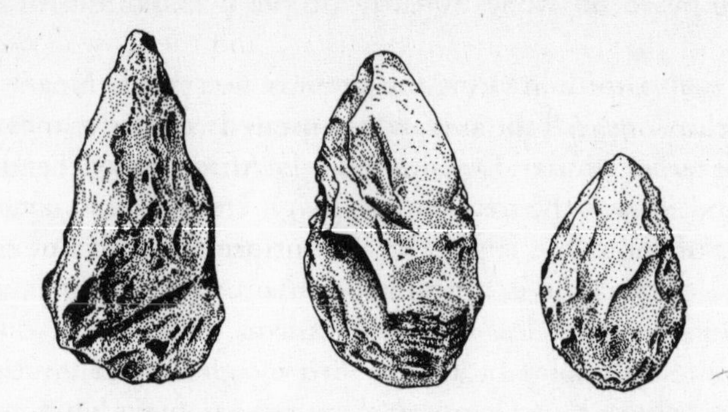

making of stone tools implies the development of language. Desmond Clark, professor emeritus of the University of California, Berkeley, has made the choicest remark about the supposed intimate connection between the two: If the hominids of that era, he said, were indeed talking with one another, and if the purpose of language had something to do with making stone tools, then we can come to only one conclusion: They must have been saying the same things over and over again for a very long time.

Whenever I see a two-year-old child opening and closing the same squeaky door time after time, or applying the loud end of a toy mallet to the same piece of wood for an entire hour, I am forced to recall that we are indeed the descendants of the people who made handaxes.

There is more to this point than might at first appear. Language, in the modern human sense, is a fantastic generator of new things. Listen carefully to someone speaking and notice the rapid sequence of totally arbitrary sounds. This phenomenon is odd enough, but even in the words we combine, the sentences we string together, and the nuances of meaning we convey, something new comes out of our mouths almost every time, something never said before in quite the same way.

This innovation and continual transmission of new meanings had no counterpart in the lithic utensils of the early Stone Age. The basic outputs of toolmaking activity varied very little over enormous spans of time. Contrast this with the late Stone Age after 50,000 years ago. As far as we know, stone was still the primary material for making tools, and certain forms manufactured tens of thousands of years before were still being made. But a new path was taken for the first time in stone technology, the path of things short-lived, temporary, impermanent.

The stone technology of this period, the Upper Paleolithic, was much more diverse regionally, and the rules of tool manufacture began to change more frequently. Fashions were born, and the sense of something permanent, embodied by the handaxe, was destroyed. Novel ways of shaping flint and other kinds of rock appeared regularly, measured in thousands of years rather than hundreds of thousands. The limits on what could be manufactured were lifted.

According to one hypothesis, this transformation occurred as new mental capacities evolved. According to another, tool kits became diverse because people began to use objects to define social differences between themselves as populations became more crowded. Archeologists are not yet cordial to one another about which hypothesis is correct. Whatever the reason, innovation shifted from second to fourth gear in short order. The social mechanisms underlying both tradition *and* the tendency to change had merged. Diversity, flexibility, mutability of behavior, became the plinths on which the human species has stood ever since.

I am in continual awe of this transition. The other day I found a slide rule made of gray plastic lying in the bottom of a carton. It amazed me that I still had one. A moment later, I was astonished to think that this helpful invention should now be so obsolete.

My Maasai friends of Olorgesailie are some of the most staunchly traditional people in the world. Yet a wizened elder who lived near my camp owned a boom box, awkwardly squirreled away under a red body cloth that barely covered him. And children herding goats sometimes pass through our camp to watch me type on my laptop. In any society, people are helplessly curious and stimulated by novelty.

How did this happen? Why did the unfailing commitment to the stone handaxe give way to flexibility and the motley burst of human activity?

3

The rift valley of East Africa is laden with age, the kind that instills a sense of permanence. Mountain peaks, deeply molded, speak of something ancient and original. Mahogany walls of lava define the boundaries of the rift, looking as if they have existed forever. The silent rains and dry winds create seasons as eternal as the human lineage. Gusting winds sweep the eroded gullies and ancient cliffs of Olorgesailie, transmitting the deep fragrance of acacia blossoms, a lingering scent of perpetuity. The branches of thorns are speckled with an equatorial snow, and the valleys fill with white petals. You get the feeling that this has happened again and again, over a timeless cycle.

Whisks of dust spin across the gullies and short-grass plains. An invisible fury collects itself into a powerful column, the speeding ghost of a dust devil. You see them again and again in a ceaseless friction between sunlight, rising heat, and dry ground. The roar and towering stature of these sedimentary creatures, breathed into life by the merest perturbation of wind, give further evidence of something primordial and perpetual.

Then there are the handaxes. Olorgesailie is one of the world's most important reserves of these wondrous symbols of Paleolithic permanence. Littered by the thousands over the ground, they capture an act of duplication so durable that the most human of all beings at that time continued to strike them from rocks for hundreds of thousands of years. The tradition that sculpted their final form was consistent with their durable source, the solid stone outcrops still visible today.

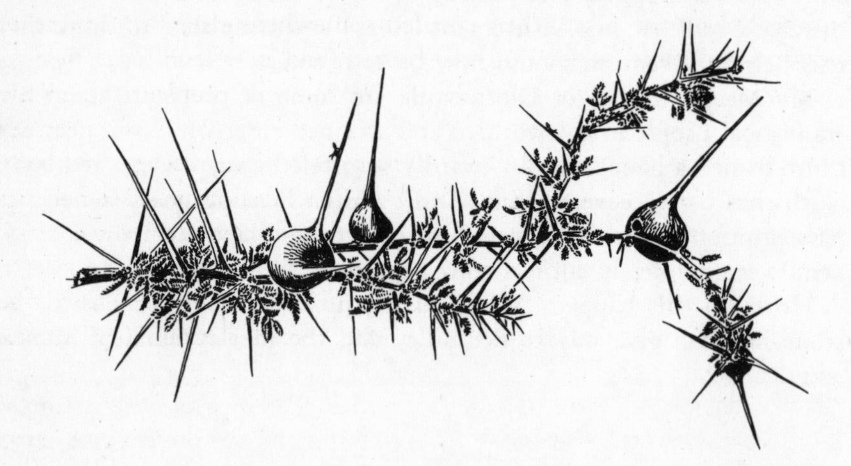

The first fact in any exploration of Olorgesailie is that a broad, quiet lake used to exist here. This fact has, however, the scent of betrayal about it. The awesome symbols of everything we deem to be immutable and permanent are actually the echoes of transience and change. At Olorgesailie, the things that incite a sense of the eternal are, in fact, abandoned fortresses, symbols of impermanence. Altogether, they create an astonishing archive of the ephemeral.

The ancient mountains and lava ridges, apparently fixed and quiescent, give rich testimony to volatile forces of change. Mount Olorgesailie was once a volcanic cone, explosive, full of fire. Modern humanity may think it has far more to lose to natural catastrophe than earlier hominids did, but the geographic risks were spread far less widely then, and catastrophes affected a far larger proportion of the species. Lives were lost in the tick of a generation, perhaps in the wave of a poisonous, volcanic cloud.

Widespread volcanic strata punctuate the geologic archive of Olorgesailie. Ash and pumice have blanketed the landscape in rare and punishing episodes, covering grass and lake, altering chemical properties and the cycle of nutrients in the existing habitats. Though the volcanoes now lie dormant and stabilized, they had a long, potent history of altering the landscape.

The rift valley floor here consists of fragmented tiles of dark trachyte rock, which form high, unyielding ridges. Stable in the present, they are proof of phenomenal spasms of change in the past. Lava erupted and oozed from fissures throughout the southern rift, spreading great layers of magma over thousands of square kilometers. As they erupted and ceased in one place, then erupted somewhere else, old panoramas were pushed aside, vegetation was buried, and new landscapes formed as the rich blacktop gradually cooled. The handaxe people occupied the region in this slowly violent time.

Between 600,000 and 660,000 years ago, tectonic movements were on the rise. Earth's crust quaked with fierce capriciousness, sending up massive layers of hardened trachyte that changed the lay of the land. Fracturing and tilting of the landscape altered the position and size of the lake. Stream channels and pathways of rain runoff were also disturbed. The water balance of the entire region surrounding Mount Olorgesailie was changed.

The dust devils, made up of tiny silica skeletons of lake diatoms, also betoken nature's transience. The lake is no longer present and never offered a consistent definition of something *original.*

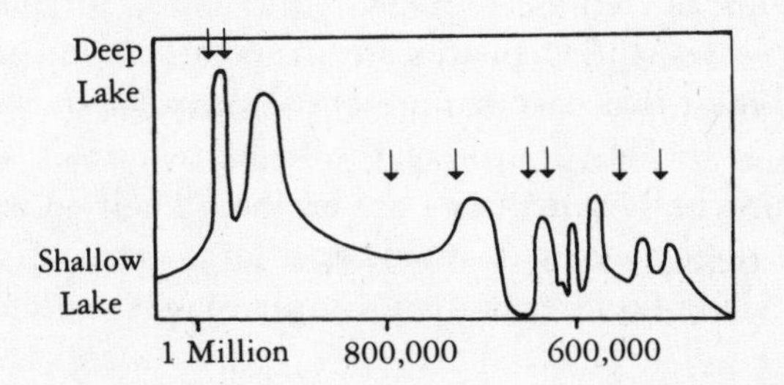

The kinds of diatoms preserved in the strata of Olorgesailie mirror the history of the ancient lake. The curve on the previous page depicts the lake's fitful rhythm of dilation and shrinkage between about 1 million and 500,000 years ago. While it indicates considerable fluctuation, certain layers—widespread bands of gravel and soil, pointed out by the arrows—reflect the lake's disappearance. The combination of diatoms and strata tell a complex story: The lake underwent a rambunctious regimen of movement in response to earthquakes and cycles of expansion and contraction indicative of climatic change.

For thousands of years at a time the lake was relatively stable. During these periods it may have seemed as if it were permanent. Then some movement in the bedrock or variation in rainfall would change its size, depth, chemistry, and the overall architecture of the terrain. On occasion, the lake dried up completely, and white powder replaced the waters; though it would return, its size and freshness were provisional.

Eventually, early humans who visited the area would have experienced the shifting environmental destiny of the southern rift. In some of the layers, hominids left their stony business cards in great abundance. In others, they left none. The handaxe makers' visits were long, but their stay was not permanent. We may wonder what effect the complexity of lake levels, climatic alterations, earthquakes, and volcanic eruptions had on the movements of these toolmakers, in and out of the region. And, ultimately, in the survival of their lineage.

As they died, the large animals of the savanna deposited bony tokens beside the old lake. From femora and crania, mandibulae and vertebrae, ulnae and humeri, we can tell who they were, the family of mammals to which they belonged, and the species to which they are assigned. These animals lived with the toolmakers and faced the same environmental kaleidoscope.

They were, in fact, very similar to the animals of the modern African plains. Remains of zebras, monkeys, antelopes, giraffes, hippopotamuses, rhinoceroses, and elephants are all in the ground at Olorgesailie, which may suggest that past and present savanna habitats are very much the same. However, about half of the fossil mammals at Olorgesailie no longer roam the savanna; they are extinct. Their council included a frighteningly huge species of monkey, a large form of zebra, at least two inordinately bulky pigs, a disturbingly massive species of elephant, a few kinds of grass-eating antelope, and a heavier-than-usual hippopotamus.

Everything we know about them indicates they were products of the Cenozoic decline, dedicated to eating grass. Their cheek teeth were large and specialized for chewing abrasive blades of grass and herbs. Their bodies were larger than their ancestors' and more massive than equivalent forms on the present savanna. These anatomical facts mean that they took on the quintessential approach to the Pleistocene plains: Eat grass, and eat it abundantly. I tend to think of them as huge lawnmowers.

The beasts inhabiting the African savanna today fall into place. The bovids—the wildebeest, gazelle, and buffalo—are the most prevalent, typically three to eight times more abundant than the equids, the Grévy's and Burchell's zebras. The suids, the warthog and the bush pig, are usually the next most abundant after the zebras. Monkeys, carnivores, rhinoceros, hippopotamus, and elephants are rarer.

At Olorgesailie, mid-Pleistocene animals had a different ranking. Between 1 million and 500,000 years ago, the lawnmower species ran a peculiar relay race in which first the zebras, then the monkeys, then the suids and hippos were the dominant animals. At the start, the most prevalent species was the Olduvai zebra (*Equus oldowayensis*), so named because its fossils were first discovered at Olduvai Gorge. Altogether, the zebras, including the three-toed horse *Hipparion* and early representatives of Grévy's and Burchell's zebras, account for nearly 50 percent of the fossils in the 1-million-year-old strata.

By about 780,000 years ago, the zebras dropped back and the giant gelada monkey, *Theropithecus oswaldi,* took their place, accounting for an amazing 50 to 90 percent of all fossils found around that time. By 600,000 years ago, the giant gelada had passed the baton. The dominant animals now were two large forms of pig known as *Metridiochoerus* and *Kolpochoerus* and two species of hippos, the extinct form, *Hippopotamus gorgops,* and the living *Hippopotamus amphibius.*

Fossils, however, offer nothing like an accurate census. The path by which living things enter the fossil record is tortuous and strange, susceptible to strong factors of preservation; the unwitting paleontologist who reads literally from the fossils strewn along the path is often blind to the pitfalls. The only message we can take away from our calculations is that the antelopes, which prevail in all East African savannas today, were never the most prevalent animals in any fossil sample from Olorgesailie, despite the diverse conditions of preservation at this site over a span of half a million years.

By the end of this time, the ruling parliament had neared its demise. By around 400,000 years ago, the entire suite of lawnmower species had died out. An era had come to an end. Ironically, this age so full of life can now be known only in bits of mineralized bone. The surviving lineages were relatives of the extinct forms; they had already evolved by the time their "grazing-club" cousins assumed dominance, but they lived in their shadow.

We tend to think of the living representatives as an open window on the distant past. Through them, we envision and define the African savanna, past and present, and resurrect the ancient savanna ecosystems of this great continent. The game park visitor is moved to contemplate time immemorial, and the student of today's migratory herds typically imagines he sees what the past was really like.

But the perspective we seek involves the succession of species over time, the longevity and extinction of lineages, including that of the hominids, who also once inhabited Kenya's rift valley. Based on the dynamism of the rift we have already encountered, we can no longer assume that Africa's savanna represents a primordial, unchanging setting—a single, original environment where human ancestors and other large animals gradually honed their modern ways. The fossil animals echo the theme I've now learned to expect from Olorgesailie: What was, is no longer.

4

A species at a time, the mid-Pleistocene grazers of Olorgesailie checked out. Why did certain animals meet their demise, while closely related species continued? This question, which follows a tantalizing track beside our pursuit of human origin, can be addressed by several closely related bodies of evidence.

During the mid-Pleistocene, the lay of the land in the southern Kenya rift, the availability of water, and the distribution of resources were sharply and repeatedly remodeled. Landscapes were transformed not only by climatic shifts but also by the more erratic schedule of earthquakes and volcanic eruptions, which suggests that the conditions of life were repeatedly altered over long periods of time.

The relay of temporary dominance among grazing animals further suggests that the surviving populations, the large mammals that now inhabit the East African savanna, may have been better able to withstand the spectrum of environmental change than those that became extinct.

On the basis of comparisons between pairs of animals—the survivors and the extinct species—the surviving lineages consistently appear to possess more flexible strategies of environmental adaptation than the extinct forms. The latter, in each instance, committed themselves to more specialized patterns of growth—larger bodies, bigger teeth—and a narrower range of foods than the survivors did. Consider the following sketches of where the contrasts lie:

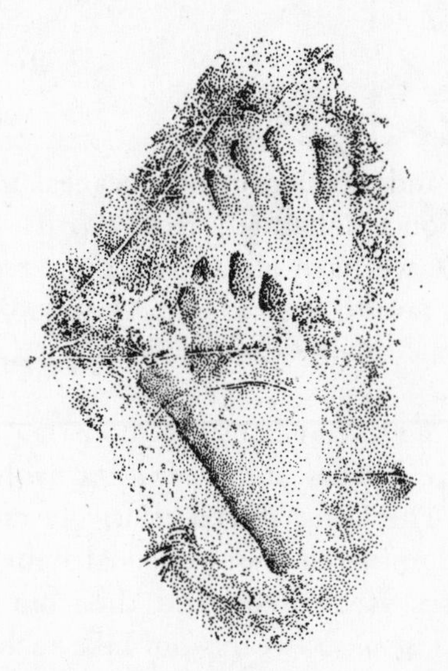

The giant gelada monkey (*Theropithecus oswaldi*) versus the common baboon (*Papio anubis*): The most abundant primate of the southern Kenya rift about 800,000 years ago was the giant gelada. It was the largest monkey ever known and, based on studies of its limb bones, the most terrestrial. Males are estimated to have weighed over 180 pounds, females about half as much. The largest males equaled the bulk of a small adult gorilla, the largest of all living primate species. It devoted its energy to growing a large body and was strongly committed to eating grass and possibly small seeds, much like its smaller living cousin (*Theropithecus gelada*), now confined to the highlands of Ethiopia.

The common baboon, which replaced the giant gelada over most of East Africa, is large for a monkey but considerably smaller than the extinct gelada, and moves with great versatility on the ground, in the trees, and in rocks, cliffs, and thornbushes. It also has an extremely diverse diet of whatever is available—blossoms, tubers, grass, even animals up to the size of a small gazelle. Although the common baboon had evolved in Africa by the late Pliocene, it did not become the main monkey of the southern Kenya rift until the giant gelada became extinct. Today, it is nearly ubiquitous in East Africa and, infamous for raiding crops, has even accommodated to the presence of humans.

Reck's elephant (*Elephas recki*) versus the modern African elephant (*Loxodonta africana*): Both *Elephas* and *Loxodonta* evolved in Africa about 4 million years ago. The latter stayed relatively rare, while the lineage of Reck's elephant dominated the scene. Although the genus *Elephas* endures today in Asia, Reck's elephant died out in Africa sometime between 600,000 and 400,000 years ago. Like each of the extinct lawnmowers, Reck's elephant was larger than its surviving counterpart. Its molars formed a high, resistant rasp, effective in dealing with huge quantities of grass and possibly other abrasive plants. It thrived in the dry, open environments associated with other super-grazers of the African plains.

Loxodonta africana, in contrast, inhabits Namibian deserts, dense forests of Zaire, and a vast variety of savannas throughout the continent. Individuals and entire herds are known to migrate vast distances, following the patterns of rainfall and changing vegetation. They can browse from trees or live on grass, sometimes switching between the two. African elephants have found ways to adapt to new circumstances and have a consummate ability to alter the habitats in which they live, occasionally uprooting trees while seeking water, significantly changing the vegetation.

The big, extinct pigs versus the warthog (*Phacochoerus aethiopicus*): The dominant pigs of the mid-Pleistocene consisted of two lineages, *Metridiochoerus hopwoodi* and the genus *Kolpochoerus,* both of which experienced increases in body and molar size indicative of a specialized, abrasive diet. The hopwood pig, a dramatic illustration of the large-lawnmower adaptation, reached the size of a small hippopotamus. By around 500,000 years ago, both lineages had met their demise, at least in southern Kenya. The warthog, the sole surviving savanna pig species, is an eclectic eater, unconfined to any one type of food. Its lineage is at least 1 million years old, although it was extremely rare in Africa until the late Pleistocene. The warthog is now the top pig of the savanna, prevalent in parks and reserves of eastern and southern Africa.

The gorgops hippo (*Hippopotamus gorgops*) versus its surviving cousin (*Hippopotamus amphibius*): The hippopotamus is the aquatic king of the savanna. Both the modern hippo and the more massive gorgops hippo inhabited the southern Kenya rift simultaneously. Both were grazers and dependent on water. The unique feature of the gorgops hippo was its large, raised eye sockets, which protruded above the surface like alien periscopes. They are thought to have been a specialization for living in water. The modern hippo is more versatile in its adjustment to both land and water. According to paleontologist John Harris, the modern *Hippopotamus* develops either more aquatic or more terrestrial habits, depending on its ecological situation. The surviving savanna hippo seems capable of responding to variations in the most critical factor of hippopotamus life, the presence of lakes and rivers. During the Pleistocene, if the environmental record of Olorgesailie is anything to go by, the size and location of watery environments underwent vast change, giving the hippos of the rift valley a rich and precarious paradox, a shifting aquascape, which the survivor was able to endure.

The short-necked giraffe (*Sivatherium*) versus the long-necked survivor (*Giraffa camelopardalis*): Around 1.5 million years ago, at least four species of savanna giraffes inhabited East Africa, including *Sivatherium.*

Only one lineage of savanna giraffe has lived to the present, the one with the greatest reach—the longest neck and the longest legs. According to traditional wisdom, giraffes were forced to adapt to eating taller and taller foliage in response to directional change in the environment. John Harris's work puts a different slant on the matter. Harris notes that each species of giraffe would have been able to reach higher foliage than most other herbivores. Their only real hindrance would have been a reduction in high foliage. According to Harris: "A long neck being capable of reaching both low and high browse, it was perhaps inevitable that one of the taller species should triumph in intrafamilial competition." The giraffe able to cover the entire range of browse, from high to low, would have been the best at buffering changes in the amount of arboreal food. In line with Harris's scenario, the giraffes with shorter necks did not enjoy the same coverage and malleability. The surviving giraffe can wrap his hungry tongue around leaves that no other ground mammal can touch, and can bend to taste even the lowest foliage.

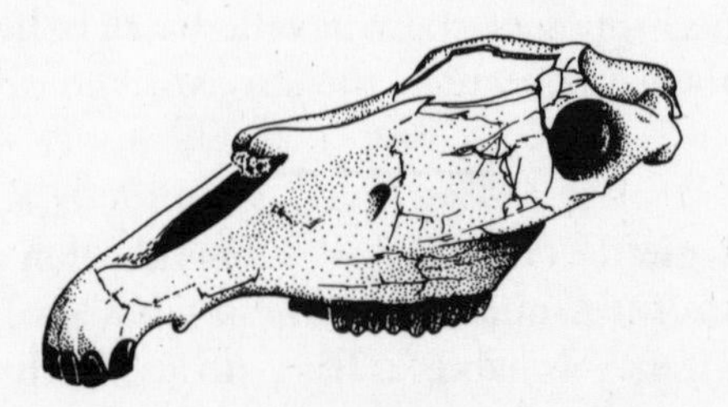

The Olduvai zebra (*Equus oldowayensis*) versus the Grévy's zebra (*Equus grevyi*): In the Olduvai zebra we witness once again the lawnmower syndrome—a big body with big teeth, indelibly committed to swathes of grass. Other than size, the only known anatomical difference between it and its closest living kin, the Grévy's zebra, is that the Olduvai zebra had an elongated, very wide snout that housed a scythe of incisors used in harvesting grass. Its muzzle could reach deep into the grass and chew off many blades per bite. The Grévy's zebra also grazes on roughage, but field studies have revealed its great ecological versatility. It easily accommodates to diverse species of grass and herbaceous plants, and when all the dry-season grass has been eaten, the Grévy's relies on browse. Females and males change their geographic range, social behavior, and grouping patterns according to variations in water and food supply. Mares and foals live by established bodies of water, while others, living away from water, group together in a wide variety of ways, evidently in response to environmental uncertainty. According to biologist Joshua Ginsberg, such shifts result from large temporal and spatial variation in the availability of resources. So in the extinct zebra we see hints of specialization in diet and life history, while close study reveals surprising variability in the life of its most closely related survivor.

In comparing living and extinct animals, we will never have complete data about the fossil species. We will never know if the Olduvai zebra had herds that varied as much as the Grévy's zebra in their organization, or if the mid-Pleistocene lawnmowers spanned a wider range of environments than the fossil record indicates. My summary is thus biased in favor of the versatility of living forms. It is, moreover, not always easy to distinguish between specialists and generalists among living species, much less to apply these labels to animals no longer around.

Despite these uncertainties, the large lawnmower species seem consistent in their anatomical and dietary commitments. Each lineage evolved a range of options limited by body size, diet, and metabolism. To some degree, their lot was cast in a way that, I believe, was overcome by their surviving relatives, who seem to possess flexible means of accommodating to variable environments. The roller coaster of habitat change, exemplified in the rift setting of southern Kenya, meant that a strong, narrowing commitment to any one tendency would ultimately prove disadvantageous.

If we could go back and wander the rift valley through a period of time, we would see a stable landscape for hundreds or even thousands of years. Food and other resources would be distributed in a particular way. The seasonal location of water would be relatively consistent. The environment would possess a certain framework into which the animals of that time would have to fit. But was the outcome determined by the survival of the fittest?

I think the answer must be no. The environment of the southern rift had neither a steady baseline nor a fixed vector of change. "The fittest" had no consistent definition, and over the long term no such definition could take root.

The dominant creatures of the mid-Pleistocene were in close touch with the main direction of environmental change during the Cenozoic—the shift to dry, open, grassy terrain. In all of their peculiarities they exuded the vital characteristics of animals fabricated by the process of environmental decline. Creatures of the dry savanna, in that kind of habitat they were certainly "the fittest." They are, however, no longer with us. To all appearances, the outcome was a matter of the survival of the generalist.

Another body of evidence hints at how habitat fluctuation affected life in the southern Kenya rift. During the mid-Pleistocene the geographic ties between species were increasingly severed. Earlier, the geography of species was relatively stable, but with the onset of larger and larger shifts in environment, populations took to longer ranges of movement, beyond their familiar confines.

The main example of this phenomenon is at Lainyamok, just southwest of Olorgesailie. This site has provided a tremendously rich fossil fauna, perhaps the only accurately dated fauna of its time period, approximately 350,000 years old. Excavations in 1984 unearthed a total of forty-two species of large mammal, every one of which can be seen today in the parks and reserves of Africa—the common baboon, the Grévy's and Burchell's zebras, the warthog, diverse antelopes, and so on. Not a single sign has been found of the suite of species that, a short time before, governed the southern rift. The lawnmower species were extinct and unknown from rift valley sites later in time. This means that their demise in East Africa occurred sometime before 350,000 years ago.

Since the diversity of animals at Lainyamok consisted entirely of living species, we might think environments of the rift had finally reached a normal state by today's standards. Antelopes were present in their expected quota, ahead of all other ungulates. Zebras, monkeys, and other savanna-dwellers also registered their typical numbers. But the odd fact about Lainyamok concerns an unexpected guest, an antelope called the blesbok, species *Damaliscus dorcas.* It was evidently limited prehistorically, as it is today, to southern Africa. Its only definite prehistoric showing in eastern Africa occurs at Lainyamok.

The surprising thing is that in its one known appearance in eastern Africa—some fifteen hundred miles north of its usual home territory—the blesbok was the most abundant antelope around. There were more of them in our fossil collections than wildebeest, Grant's gazelle, impala, or any other single species. What was it doing at Lainyamok? Why was it living in a strange land, independent of any of its southern neighbors, mixed up with animals quite at home in eastern Africa, yet more abundant than any of them?

The lesson of the blesbok goes beyond the rift valley. It is something that paleontologists have deduced in later Pleistocene habitats of both Europe and North America, but it has been largely unnoticed in Africa. It has to do with the disruption of links between species.

We often think of species that live together as a kind of integrated unit. Certain species hang out together because of some common attraction and may depend on each other's presence in the ecosystem. At Olorgesailie, a sea of grass may have been what brought together the coterie of big lawnmower species during the mid-Pleistocene. For whatever reason, a certain set of species will assemble and live together for a long time.

Over the past 500,000 years, however, animal populations seem to have developed a stronger sense of independence. Populations of many different species expanded and contracted to their own beat, moving in and out of regions as though exploring some intricate maze, or reacting to some erratic rhythm.

In northern latitudes, evidence of this phenomenon is best found in fossil assemblages younger than 15,000 years. Paleobotanists working in North America and Europe have measured different rates of migration for different species of plants. As glaciers retreated, poplars and spruce trees quickly invaded the barren terrain, while white pine and hemlock migrated much more slowly. Different species moved according to their specific life requirements. Populations of plants and animals failed or succeeded depending on their migratory speed and their level of tolerance to new climatic regimes. Communities of animals and plants came together and broke apart. Over time, new and unusual species groupings appeared for a period, until the climate altered again.

Changes in the dominant animal—zebras at one time, giant geladas at another, pigs still later—tell us that the mix of species occupying the Olorgesailie region was subject to occasional pervasive change. The blesbok of Lainyamok goes farther and suggests that the geographic ranges of some species were undergoing independent expansion, contraction, and displacement. And this was beginning to take place in sub-Saharan Africa by at least 350,000 years ago.

Over the past 700,000 years, therefore, two curious things happened to the animals of the southern Kenya rift. First, extinction rates accelerated among the previously dominant species. These animals, all bound to a particular way of life, seem to have been outsurvived by more versatile species. Second, the ties linking particular animals to particular regions and to one another became much looser. These two aspects were not related to any specific directional change in habitat. Rather, they coincided with an increase in environmental variability, suggesting a curious hypothesis: During the unsettled Pleistocene of the southern Kenya rift, the migrants and the generalists were favored over all other creatures in the region.

5

One's eye is slowly apprenticed to the dynamics of the rift. Its fractures and relief assume new significance. Rock begins to liberate a fantastic history. Traces left by the dead testify to the birth and culmination of lineages of living beings. In this newly encountered drama, the rift valley becomes a very special place. But is the rift an exception? Was its turbulent saga of land and climate limited to its magnificent lava boundaries? Was the rift merely a pocket of instability?

Nature has deposited its lengthy archive wherever there are places to gather the fallout of the past. Ocean basins, highland plateaus, and dark caves all have their own histories to tell. If we search other mirrors of worldwide climate and localized habitats, we will see that sweeping signals of habitat alteration were integral to the era and the specific milieu in which the human lineage evolved. The long-term extremes of life conditions affected the survival of species, the present shape of the planet's biota, and our own involvement in it.

Innumerable jagged lines could express our most important point:

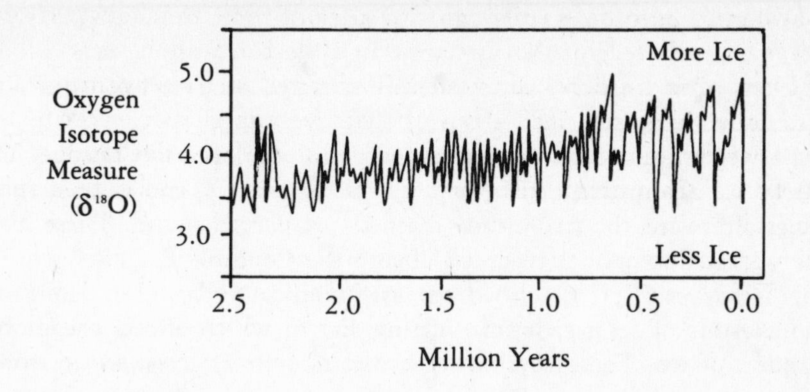

Within the past 1 million years, the strength of environmental fluctuation increased once more. Climatic disturbances, especially over the past 700,000 years, took on an intensity eclipsing that of any previous era.

The above example was chosen because it can be compared to other data we've examined. It shows the measured oscillation in oxygen isotopes, based on forams drilled from the ocean bottom. The forams reflect the glacial pulse, in this case the rhythmic buildup and melting of ice worldwide over the past 2.5 million years. An upward twitch of the line indicates the growth of massive sheets of ice at high latitudes and on the tallest mountains, even on the equator. A downward turn means that water locked up in glaciers was returned to the oceans and atmosphere as warmer temperatures prevailed.

Vigorous oscillations occurred between 3 and 2 million years ago. In the middle of that interval, the intensity of fluctuation—the amount of zig or zag in the curve—reached 1.0 to 1.2 parts per mil. Another time of marked change was around 2 million years ago. Important events in the evolution of human beings coincided with these periods of intensified climatic oscillation.

Based on the curve above, fairly low-level oscillations filled the span 1.9 to 1.5 million years ago. During that time, the most notable shifts in oxygen isotope ratio fell to between 0.3 and 0.7 parts per mil, averaging about 0.5 parts per mil. Both the mean and the entire range of environmental change were considerably smaller than in the previous 1 million years.

After 1.5 million, the flickering of the isotope curve changes once again. The amplitude of change rises above 0.5 parts per mil on a consistent basis, with a maximum range of 0.9 parts per mil.

Around 1.2 million years ago, a change of more than 1.0 parts per mil was registered for the first time in almost a million years. Modifications of at least 0.9 parts per mil occurred at least five times immediately thereafter, until about 700,000 years ago.

Between 700,000 and 600,000 years ago, vacillation in isotopes and glacial ice again extended beyond 1.0 parts per mil, and at least three times afterward the magnitude reached 1.5 parts per mil. These were the widest fluctuations recorded during the Cenozoic Era.

In Chapter III, I described the astronomical cycles that determine the amount of solar radiation hitting Earth, which affects the global buildup of ice. These rhythms, the Milankovitch cycles, are mirrored in our isotope diagram. The zigzag line shows an important shift in the rate of climatic oscillation, corresponding to a change in the main orbital influence on world climate.

This change is best illustrated by examining two parts of the oxygen curve. Consider the period between 1.9 and 1.5 million years. Twenty major oscillations are apparent in this 400,000-year interval. A major alteration in the global volume of ice occurred about every 20,000 years. A complete cycle—say from one glacial peak to another—occurred about every 40,000 years. This period reflects the 41,000-year shift in Earth's tilt—a cycle believed to be mainly responsible for the rate of ice ebb and flow over the span of human origin, and possibly over the entire Cenozoic, up until about 1.2 million years ago.

At that time, the 41,000-year signal began to be overshadowed. You can see its decay in the curve between 1.2 million and 700,000 years. The climatic pendulum began to make a wider arc, taking longer periods of time. The bold, vigorous pulse of the 100,000-year rhythm became louder and louder, corresponding to the cycle of change in the shape of Earth's orbit around the sun. After 700,000 years ago, the oxygen isotope curve prominently displays the slower, more powerful throb of this 100,000-year cycle.

The cause of this rhythmic change has been a target of scientific research, but no consensus has been reached. One factor or another has been proposed: the rise of vast highlands, such as the Tibetan plateau; a steepening of the heat gradient from the equator to the poles; the overall decline in forest vegetation. All of these changes indeed took place during the late Cenozoic, and doubtless altered the way in which solar radiation was translated into weather patterns across the globe.

Whatever the precise set of causes, the planet was poised to descend into a dramatic weather moodiness that has persisted to the present. Sharper, more pronounced oscillations sculpted the next 700,000 years of our evolutionary history.

The span covered by the isotope curve, particularly the last 1 million years, is popularly known as the Ice Age. This suggests that the glacial covering of northern continents was the paramount event of the period. But the trend toward more ice was broken repeatedly and decisively by larger degrees of melting than at any time in the previous 1.5 million years. Although the ice epochs were longer than the periods of meltdown, the climatic pendulum's wide path encompassed both extreme ice and extreme warmth.

Over the past 500,000 years, there have been at least three major intervals in which the volume of ice fell below that of today, which is also an interglacial period. Each of these interregnums of warmth was succeeded by a reign of intense cold as ice built up on land. The great range of oscillation was the principal environmental theme of the past 1 million years.

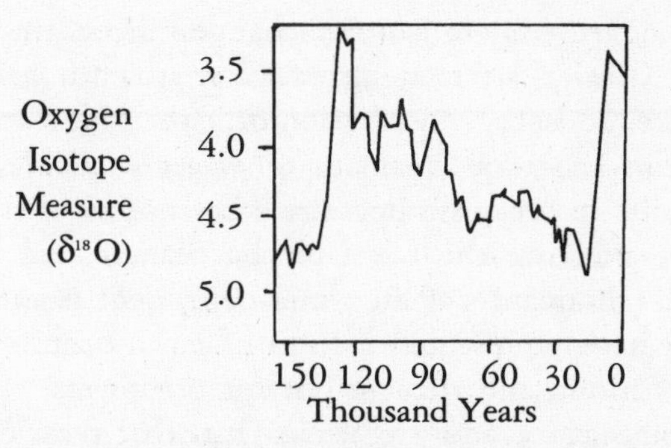

Like an ominous seismic report, the zigzag line above conveys the tremor in the oxygen isotope curve between 128,000 years and the present. If we narrow in on the span between 128,000 to 72,000 years ago—a time of interglacial warmth sandwiched between glacial periods—we find numerous changes in the curve, reflecting shifts in the direction of global ice buildup. No fewer than ten reversals can be discerned over a period of 56,000 years. Some of these shifts persisted for a couple of thousand years: others lasted almost 10,000.

What makes this particular period stand out is the amplitude of change. The range of fluctuation was considerably higher than it was during either of the lengthy glacial periods immediately before and after. The rapid removal of ice at the outset of this period seems to have been faster and more complete than the deglaciation 11,000 years ago that led to the present regime of global climate. I find it interesting—and, as we will see, more than a coincidence—that the appearance of modern *Homo sapiens* and the oldest developments in symbolic activity, which are the foundation of modern cultural behavior, occurred during this span of high amplitude change beginning about 130,000 years ago. This period included not only a full rebound of the dominant 100,000-year cycle, but also significant reversals within the less dominant cycles that affect the hydrology of the planet.

Experts on ice-age climate suggest that the late Pleistocene was characterized by two stable states—glacial and interglacial—with short transitions between the two. The details of the ice-warmth curve, however, depict innumerable transitions of varying magnitude, leaving room to doubt whether the terms "stability" and "state" accurately convey the environmental conditions faced by species known to endure over hundreds of thousands of years.

The long, jagged edge of isotopic change mirrors the plumbing system of Earth. Glaciers are made up of water stolen from the oceans and the atmosphere. So isotopic oscillation over the long haul denotes large irregularities in the total quantity of water available to organisms. Long-term shifts in ocean isotopes are a bellwether of the most important variables affecting the biota of the planet—the distribution of water and the abundance of all water-dependent resources. The fluctuations may have represented a kind of equilibrium—a rebounding between ice buildup and decay—but for organisms, the implications were more complicated and precarious than that term connotes.

Change in sea level was a direct effect. During maximum glacial times, almost one third of Earth's land surface was covered with ice. As a result, ocean levels dropped as much as one hundred meters. New land was exposed on the continental shelves. New links between continents and islands were established. Generations of animal and plant populations responded to the expanded coasts and corridors of movement. Over time spans significant to evolution, many things changed. Ice volumes, habitable land, and the avenues connecting one region to another all proved ephemeral.

Astronomical cycles determine the amount of solar radiation that

Earth receives. Yet the intricate blend of these cycles is merely the first step in the growth and decay of glaciers. Ice-age effects are caused by a complex intersection between ocean, ice, winds, and weather, coupled with volcanic forces and the traumatic uplift and downthrust of Earth's crust.

A fantastic array of local and global factors joined together, some amplifying and some conflicting with the tendency to add or subtract glacial ice. The various effects are hard to understand because they were subject to complex feedbacks and thresholds—what scientists call *non-linear interactions.* As ice sheets grew, they reflected solar radiation back into space. This reduced the effective amount of solar radiation on Earth, further lowering the global temperature. The spread or contraction of forests modulated the amount of moisture produced by plant transpiration, and this, in turn, affected the amount of water available to convert to continental ice. One perturbation returned the effect of another in an astonishing dance between Earth processes and living things.

As a result of these interactions, environmental change was expressed in different ways in different regions of the world, and each ice age or meltdown looked somewhat different from the one before. Each major zig or zag in the isotope proxy of global change is a signal of new topographies and mosaics of vegetation never seen before. Each major shift in global climate created unfamiliar rearrangements of water, food, and other resources.

In Devil's Hole, Nevada, a calcite core exposes the past 566,000 years of variation in carbon and oxygen isotopes. The *oxygen* curve tells about global ice variation; it coincides with the record of the deep ocean forams, though it is curiously at odds with the expected ages of the Milankovitch orbital cycles. The *carbon* curve, on the other hand, indicates both the vegetation that existed in the region and the amount of carbon in the atmosphere. The latter, like the oxygen curve, is a mirror of climatic change on a large scale. The strange thing is that the oxygen curve of global ice change consistently lags behind, by about 7,000 years, fluctuations in the carbon curve. It is unclear why, but it, too, suggests that either cycles of deforestation-recovery or changes in global carbon were out of sync with the cycles of ice buildup and decay in the latter third of the Pleistocene. Whatever the exact reason, the range of variation in both the oxygen and carbon curves increased over the past 500,000 years, adding another record of the ever-widening range of environmental vacillation during the Pleistocene.

The interaction of volcanic eruptions, earth movements, orbital cycles, and weather patterns had many unexpected results. According to one recent proposal, the growth of glaciers placed an enormous burden on portions of Earth's crust, triggering volcanic eruptions far away. As the volcanoes discharged large quantities of particles known as aerosols, sunlight was blocked and the growth of glaciers was enhanced even further. The two factors were coupled in a cataract of change. Michael Rampino and Stephen Self have thus suggested that the eruption of Mount Toba in Southeast Asia over 70,000 years ago pushed an initial glacial advance in the Northern Hemisphere over the edge, and in a rapid plunge, the last major ice age was born.

In the southern Kenya rift, volcanic activity, earthquakes, and climatic change joined in a cascade of oscillations during the mid-Pleistocene. Change in the shape of the landscape and in the regimes of moisture and drought were coupled. It has always irritated me, from a scientific point of view, that the effects of climate, tectonics, and volcanism on the changing habitats of Olorgesailie are nearly impossible to untangle. But this may be the most interesting finding; this may be exactly how environmental change happened in the Great Rift and elsewhere—a conspiracy of inseparable causes.

Lake-level change, habitat variation, and faulting all occurred at a faster pace at Olorgesailie between 700,000 and 600,000 years ago. This was the time when fluctuations of oxygen isotopes in the deep sea and of ice over the planet's surface became larger. Such coincidences—widespread change in the mode of fluctuation—hint at tantalizing connections over great distances in the processes of Earth, between land and sea, high latitudes and the tropics.

In northern regions, the periodic buildup of huge glaciers transformed the landscape. Ice engraved great valleys, unearthed tons of rubble, squeezed and moved entire zones of vegetation, created tundras and steppes unlike those of the interglacials, left behind new lakes and other new features of the landscape. The glaciers withdrew. With each new advance, the transformation would begin again.

Eighteen thousand years ago, two ice sheets spread across Canada into what is now the United States, south of the Great Lakes. Glacial ice blanketed most of Great Britain, covered northern Russia, and extended from the Alps and other major highlands, possibly over the Tibetan plateau. Arctic tundra, so-called "polar deserts," extended southward, skirting into Italy, encompassing much of France and parts of the Iberian Peninsula. Cold steppes ranged from central China to eastern Europe. Forests, deciduous woodlands, and other vegetation seen in the temperate zones during interglacial times were compressed, forced to the south in complicated retreats, even eliminated along certain longitudes of movement.

During interglacials, the winds of change were reversed. Forests were regenerated in areas previously occupied by cold plains of windblown dust. Hippopotamuses lived in the British Isles. Displacements of fauna and flora were pervasive. It wasn't a matter of replacing a few key species. Entire biomes were reconfigured.

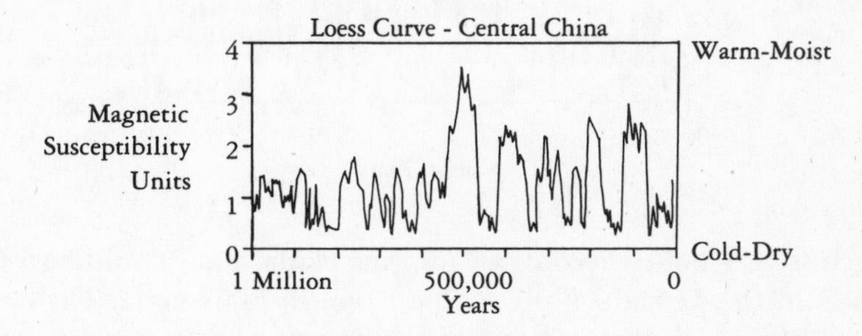

Over 40 percent of China's arable land is now blanketed by a thick sequence of wind-borne silt and sand known as loess. Formed over the past 2.4 million years, the Chinese loess record contains at least nineteen layers of dust, interrupted by an equal number of soils. The two kinds of strata, soils and loess, reflect large fluctuations in vegetation and dust. Great aridity and powerful dust storms created the loess-covered landscapes. Vegetation was sparse, dominated by grass, shrubs, and weeds. At times, near-desert conditions prevailed. The soils, on the other hand, were created under warmer, humid conditions conducive to open woodlands or dense vegetation. Some places, like Lanzhou and Xifeng in north-central China, alternated between cold steppe and warm forest.

The loess-soil deposits show that wholesale changes in plant, water, and other resources occurred repeatedly in eastern Asia during the Pleistocene. Early human populations arrived in this part of the world sometime before 1 million years ago. Over tens of thousands of years, epochs of tremendous disruption occurred, experienced not by any single individual or group, but by the lineages to which they belonged.

La Grande Pile is the name of a twenty-meter-deep peat bog in eastern France. Preserved in it for the past 140,000 years is a continuous archive of pollen. In the early part of this span (130,000 to 70,000 years ago), seven major inversions in vegetation occurred. The habitat switched back and forth between temperate forest and subarctic habitat, frigid grassland and pine-spruce-birch taiga. In certain instances, the conversion from subarctic to temperate habitat happened within 150 years, very rapid even by modern standards. Similar swings in vegetation are recorded in the peat bog of Les Echets near Lyon.

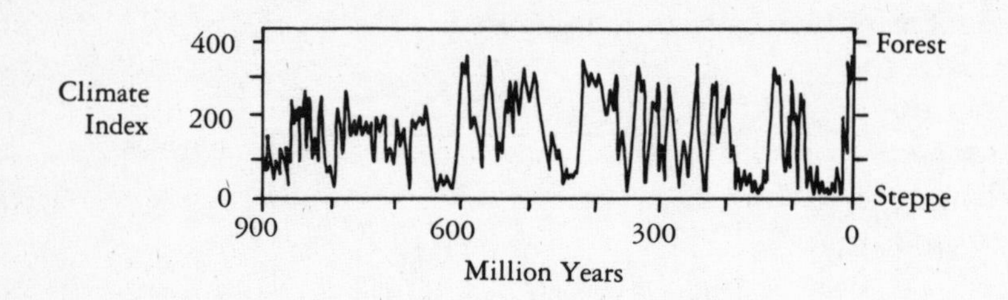

A much longer pollen record, an imprint of the past 1 million years, comes from the Tenaghi Phillipon peat bog in Macedonia. The curve, depicted above, testifies to intricate sweeps of climate, again with a change in the cycle just after 700,000 years ago. In this case, the variations ranged between closed forest and open steppe, reflecting shifts in precipitation.

What about latitudes closer to the equator? What was happening in the tropics, especially in the deepest wellspring of human ancestry, tropical Africa, where human forebears still flourished?

Variations in solar radiation, winds, and moisture were worldwide in effect, and certain areas of the tropics were hotbeds of volcanic eruptions and earthquakes. Tropical monsoons and jet streams were subject to strong alterations. The grip of climatic change was felt over most of Africa.

At least four lines of evidence demonstrate high amplitudes of habitat change over tropical Africa during the Pleistocene. The first comes from examining deep ocean cores drilled off the northwest coast of Africa. The cores preserve several markers of continental habitat—windblown quartz, terrestrial clays, and freshwater diatoms. Variation in the quantity of windblown quartz shows that multiple shifts between vegetated and desert conditions occurred in northern Africa. Terrestrial clays transported by rivers indicate that rainfall on the northwestern side of the continent varied with increasing intensity during the Pleistocene. And the presence of freshwater diatoms laid down in lakes signifies large-scale cycles of drought, which left lake beds open to erosion.

The diatoms were blown over long distances, and their increasing variability in the ocean cores over time further demonstrates an erratic chronicle of moisture and drought. The most intriguing thing is the timing of the diatom cycles, seen in a continuous core representing the past 260,000 years. The abundance of windblown diatoms rose and fell over spans approximately 23,000 years long. This frequency of change matches the precessional cycle of Earth's orbit, Indeed, other evidence also suggests that the dominant oscillation in African moisture was of this length, in contrast with the strong 100,000-year cycle of glacial growth and decay in higher latitudes.

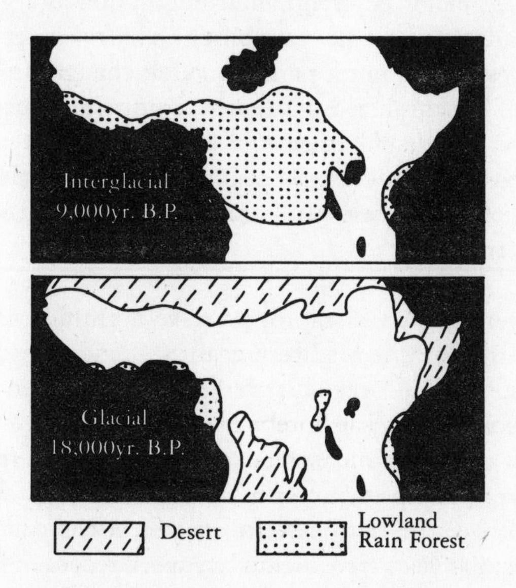

The maps above depict a second line of evidence, based on the geologic record of lake levels and fossil pollen. The past 18,000 years encompass the last major arid-moist cycle throughout Africa, when vegetation was greatly altered. This coincided with the last glacial cycle, when an enormous volume of ice expanded across the northern latitudes and then retreated. By about 10,000 years ago, the current epoch of warm, interglacial climate was under way.

As the maps show, Africa was extremely arid 18,000 years ago. Sea temperatures in the equatorial Atlantic had decreased by 4° to 5°C, greatly reducing the evaporation of water that feeds the southwesterly monsoon, which controls rainfall in western and central Africa. Most lakes had dwindled in size. Desert had spread at the expense of grasslands. Grasslands had penetrated areas of lowland forest. In the face of aridity, the tropical forests retreated to small, distant patches.

Around 12,500 years ago, equatorial forests began to expand again. Lakes overflowed their old boundaries. Between 9,000 and 6,000 years ago, even the Sahara and Namib deserts were sprinkled with wetlands and lakes. Sand deserts all over the continent contracted; grassy and wooded savannas replaced arid wastelands; lowland forests invaded what had once been dry savanna. The equatorial rain forest stretched beyond even the broad belt of central and west African forests of more recent times.

Using this last major environmental fluctuation as a guide, we learn something about the extreme variability in African settings over the past 700,000 years. It is not a perfect model; changes in moisture were not always well synchronized across the continent. But it shows that swings in moisture, displacements of vegetation, and remodelings of terrain encompassed vast areas, occasionally affecting the entire continent. Habitat conversion was not confined to high latitudes or to mere pockets in the tropics.

Thin black layers known as sapropels have accumulated over time in the bottom of the eastern Mediterranean Sea, and they lead to a third line of evidence. These organic-rich layers form when the Nile discharges an especially heavy flow of water into the Mediterranean, which implies the fall of strong monsoonal rains throughout the river's enormous source area. The presence of a sapropel indicates an extreme volume of rainfall over the northeastern quarter of the continent.

The past 464,000 years have yielded a superb record of sapropels, studied in the 1980s by French researcher Martine Rossignol-Strick. Using a mathematical model of orbital shifts in solar radiation, Rossignol-Strick computed a monsoon index precisely correlated with the occurrence of sapropels. The extremely arid period around 18,000 years ago correlates with a low monsoon index; a high index corresponds with the sapropel of much wetter times beginning around 10,000 years ago.

The most elegant item in the study is the proof that this last dry-humid cycle was not a freak occurrence. The change from arid to moist was certainly impressive, but by Rossignol-Strick's calculations, it was by no means the most extreme monsoonal fluctuation registered over the past 464,000 years. On twelve earlier occasions, the monsoon index surpassed that of the very moist phase around 10,000 years ago. And on eleven other occasions, the monsoon index dipped even lower than calculated for the last glacial maximum, when deserts expanded and forests contracted significantly over the continent.

The sapropels and the strongest monsoons appeared approximately every 21,000 to 25,000 years. Thus the sapropels also coincide with the 23,000-year precession in Earth's orbit, the cycle of interaction between annual seasons and the Earth-sun distance, which confirms the vital influence the cycle has had over Africa's trade winds and monsoon circulation. Tropical Africa adopted its own pulse of alteration, which was fast and intricate in its effect on the equatorial biota.

There is a fourth and final line of evidence of environmental change in the African Pleistocene: nearly identical or closely related forest plants and animals are disconnected by vast areas of savanna. Certain forests register very high levels of biodiversity, while the areas between them contain far fewer species. The former zones, the refugia, are small pockets of habitat believed to have served as centers of evolution. The organisms that evolved and still live there are referred to as endemic species, indigenous to the places where they are found. Many different kinds of palms and orchids, butterflies and monkeys, and other organisms evolved in these refugia, while the intervening areas harbor animals and plants evolved mainly from the endemic species.

Thus parts of Cameroon, enclaves of the Zaire basin, and other small regions were havens for rain-forest animals and plants. The current separation between them attests to vast migrations of forest and other kinds of habitat. At times in the past, the great rain forests of equatorial Africa shriveled to a few small pools. At other times, they occupied a great expanse. Highland forests of Cameroon in the far western part of the continent were connected to the highland forests of East Africa, as similarities of species between the two regions indicate. This confirms what we already suspected from pollen and lake-level records of the past 18,000 years—the tremendous dynamism of African environments over geologic time.

It also gives us a rather unsettling impression of tropical rain forests, long considered the icon of environmental stability. Let us take a brief detour to the immense Amazon rain forest, even though there is no evidence of human occupation of South America prior to very late in the Pleistocene. Amazonia is considered the epitome of habitat stability, an ancient place largely undisturbed before human interference. It has often been construed to represent something original and immutable in the natural environments of the planet. Yet, great as our desire is to believe in something changeless, this view of the Amazon cannot hold that sacred place.

The great forests of the region are a difficult place in which to study geology; there is not much in the way of erosion, so exposures of old sediments can hardly be seen anywhere in the huge fertile basin. Looking into the small windows of sediment that exist along riverbanks, geologists have nonetheless claimed that subtle signs of sand deserts lie immediately beneath the present forest floor. Others have reported evidence of extensive floodplain deposits beneath the lowland forests of Brazil. All these facts suggest that the great rain forest has been disturbed over wide areas. Some rebels have gone so far as to speak of it as a "pioneer community," younger than generally believed and neither permanent nor stable, at least on the time scales significant to evolution.

Widely accepted in Africa, evidence of forest refugia has proved controversial in South America. The extent of change in forest size has yet to be measured, but even the most cautious detractors admit that, based on meager clues, these great rain forests were disrupted prior to human presence.

If we turn to the nearest sites where long environmental records can be studied, it is clear that the tropical zone of South America experienced wide changes in habitat during the Pleistocene. In the highlands adjacent to the rain forests, a temperature decrease of up to 9°C during the last glacial maximum caused a major drop in rainfall. The best lowland record shows that the climate was often dry north of the Orinoco River during the late Pleistocene. Although it was originally thought to have been a forest refuge during dry eras, we know now that this cannot be true. When forest withdrawal occurred, it may actually have been more extensive than even the refugia proponents have suggested.

Ocean cores and sapropels near Africa mirror the transient levels of rainfall that supported forests, savannas, and deserts. While the data for South America are less secure, the few long environmental records we have indicate large alterations over the past tens of thousands of years. The third major band of tropical rain forest is in Southeast Asia. Although few long-term data have been reported for this part of the world, changes in sea level are known to have greatly affected the equatorial lowlands of the region. Periodic rises and falls in the coastline, synchronized with glacial cycles, caused potentially massive alterations to island and peninsular landscapes and to the lowland forests on them.

So the greatest icons of environmental stability we possess—the tropical rain forests—were anything but stable. Susceptible to repeated natural expansions and contractions, they echoed the ephemeral nature of moisture and temperature regimes at low latitudes. In Africa, if not elsewhere, these alterations were truly profound. The environmental records of the African continent are all in agreement: They support not the idea of natural stability but a history of deep disturbance during the epoch in which the present human species originated.

How long did it take to complete one grand excursion from forest to steppe or grassland to desert? How fast were the changes from dry to moist, glacial to interglacial? Environmental records of the past 260,000 years offer precise data. In Scotland around 10,000 years ago, an ice cap over sixty miles long took only 800 years to grow and melt again. Continuous pollen records from France, the Black Sea, and Greece all indicate that large shifts in Pleistocene vegetation—vacillations between cold steppes and moist forests—occurred in less than a century, occasionally on the scale of a decade. The largest shifts were not very frequent, but they were quick and led to hundreds or a few thousand years of relative stability.

A superb record of lake sediments from central Canada, about 4,500 years old, reveals that open tundra was transformed into a dense forest of black spruce in about 150 years. Cave deposits in northeastern Iowa indicate that sometime between 5,900 and 3,600 years ago, the invasion of forest by prairie was completed in less than 100 years. When vast remodelings of landscapes occurred, they often took place with surprising speed. Every available sign indicates that this pattern prevailed throughout at least the latter half of the Pleistocene.

In tremendous feats of engineering, long cores of ice have been drilled from Antarctica and Greenland, providing nearly continuous data about the climates of the past few hundred thousand years. One of the Greenland ice cores discloses a stunningly erratic environmental twitch, which matches the overall pattern of the deep ocean record for the last 115,000 years. The researchers in charge consider this discovery a climatic "flickering switch," which developed at irregular intervals and included intense, very rapid climate change. Shifts between warm and cold were made up of many high-amplitude oscillations in temperature, ice, and atmospheric dust. The swings were sometimes extreme, bridging glacial and interglacial conditions in a single leap.

Scientists disagree about how to interpret the Greenland data for certain time periods. The "flickering" does not show up, for example, in the deep ocean record between 128,000 and 118,000 years ago, leading some researchers to consider this a relatively stable span. Nonetheless, numerous climate and vegetation records show that large environmental shifts took place at the beginning and end of this span in 150 years or less. According to the Greenland data, another dramatic change 11,660 years ago was completed in a decade or less, but a study of pine-tree rings shows that it took place in Europe in about a century.

Despite these discrepancies, dramatic alterations between cold and warm, steppe and forest, glacial and interglacial, occurred time and again in the late phases of humanity's descent, and when they happened, the change was measured on a scale from decades up to a few centuries. Our view of nature consistently focuses on stability and equilibrium. While certain climatic states were repeated, we begin to wonder what all the oscillation was about. What was its effect on the evolutionary origins of humankind and the species around us?

6

Major biomes—savanna, forest, and desert—moved across the terrain in complex relays. Water and food—the key resources of living organisms—appeared and disappeared over time. The environmental factors susceptible to repeated, large-scale change over the past 1 million years were the very ones around which animals and plants organize their competitions and approaches to life—social grouping, the search for mates, the risks of predation, the means of survival, and reproductive success.

According to a huge literature on the subject, the phenomenon of territoriality is indelibly linked to the size and spacing of resources. Many animals' feeding and social life are governed by whether territories can be defended. Fruits sought by the forest tamarin monkey, nectar by the golden-winged sunbird, and seeds by the tree squirrel, are all in small defensible patches. Feeding parties in each of these species are also small and separated by strict boundaries. In red-winged blackbirds and many other vertebrate species, females inspect and select nesting sites on the basis of the food supply within territories won by males. Large expanses of grass or foliage, on the other hand, are impossible for any animal to control; species relying on these resources tend to feed in large herds or bands.

These factors must have varied greatly on local, regional, and continental scales during the Pleistocene, so responses to variability—the need to follow resources or to expand and contract the home range—were important in the evolution of many territorial species. For example, the California sanderling studied by biologist John Myers and his colleagues has been found to defend territories as large as possible, mainly as a means of buffering changes in food supply. The sanderling persists in this strategy even though each territory becomes prone to intrusion by competing males.

The home range of African colobus monkeys is also controlled by food. The black-and-white colobus feeds on a dense, evenly distributed supply of leaves and requires only fifteen hectares of forest foliage to support itself. The red colobus, on the other hand, requires a range more than four times larger to allow for its more widely spaced, energy-packed diet of fruit, flowers, and shoots. Since forest size governs the supplies of these foods, Pleistocene oscillations must have exerted a considerable influence on the number of home ranges, population density, and survival approaches of these two species.

Periodic shifts in the size of forests and grasslands created opportunities for variety in diet. Fruits, seeds, insects, and other rich food sources were especially susceptible to modification. Change in their distribution affected energy, metabolism, social life, mating, and all the other outcomes and determinants of natural selection. By causing population size to change, alterations in habitat area would have also changed the competition between individuals or populations at different times.

Animals do not search for food at random. In many birds and mammals, individuals take a characteristic path between separate feeding places. Animals learn a lot about their environments and develop efficient patterns of movement and use of resources. In creatures as diverse as shore crabs, wasps, lizards, mice, and lions, individuals often drop the search for particular food items and opt for the more available. In bees, primates, frogs, reptiles, and numerous birds, animals seem to identify the "best" plants to feed on, defined by nutritional value and the time it takes to locate the food. We can only imagine how these definitions changed and how the mental process of defining "best" evolved in the Pleistocene.

When animals forage, they either return to where food was last found or shift to a different locale. The success of this choice depends on the dispersion of food. If it becomes altered in any way, animals may have to change their tactics. Environmental variability thus encourages ways of finding out about food dispersion and acting on it. Learning from foraging successes and failures, remembering food sources visited over a lifetime, and deducing potentially new sources are the faculties that favor survival in an unstable setting. The more life-threatening the shift in terrain and vegetation, the more valuable behavioral flexibility becomes.

During the Pleistocene, North American deer mice (*Peromyscus maniculatus*) had to survive highly varying conditions over seasonal to geologic scales of time. Biologist Lincoln Gray's study of this species illustrates how diet breadth may be altered by a fluctuating food supply. Gray's experiments show that deer mice are adept at learning and responding to variability in environmental conditions. If young mice are fed a constant, predictable diet, as adults they specialize on a smaller range of foods. If they experience a less predictable, fluctuating food supply, they grow up to choose a significantly more generalized diet. These mice have developed a tie between cognition and food, enabling them to respond to contingencies in their habitats.

Innumerable studies have shown that resources affect not only how animals search for food but also their ways of grouping together. John Crook's pioneering research on weaver birds, a very diverse group of African and Asian finches, was among the first to explore the way in which sociality varies with ecological conditions. Each species of weaver bird was found to mate either monogamously or polygamously, to feed alone or in large flocks, to eat mainly insects or mainly seeds, all depending on whether it inhabited forest or savanna. Peter Jarman's later study of seventy-four species of African herbivores demonstrated more refined links between mating and habitat. In antelopes, the type of environment—forest, woodland, grassland—shows a close connection with mating behavior, method of predator avoidance, and diet.

More recent studies have further explored the principles of interaction between animals and habitats. An animal eats, travels, and lives alone or with others of its species, depending on the spread of critical items like water, food, and places of safety. The distribution of resources greatly affects the intensity and kinds of interactions between individuals, partly because it determines when and where individuals seeking the same resources meet.

If you are a large, long-lived, social creature, you need to know a lot about your habitat. Where are the most abundant food sources? Where are predators likely to lurk? How will other individuals respond to the same resources? Where are mates likely to be found? Social animals are good at solving these basic questions and adjust their behaviors according to the setting.

Two factors in particular—food abundance and distribution, and predator size and techniques of avoidance—exert great influence on social behavior and group organization. It is therefore reasonable to conclude that wide fluctuations in the size of biomes and the qualities of habitats have deeply influenced the evolution of animal societies.

In primates, the strength of female social bonding and the frequency of female movement between social groups depend on food distribution. Where food occurs in easily defended patches, females often have strong bonds of cooperation and tend to stay in the group in which they were born; this occurs in both baboons and langur monkeys. Where food is evenly distributed throughout the environment, as in the habitats of gorillas, females tend to leave their birth groups, and social bonds in their new groups appear to be less strong; there is less grooming and feeding together. During the Pleistocene, the spatial mosaic of food may have sometimes remained intact as habitats shifted their boundaries. At other times, it seems likely that plant-food distributions were drastically altered, requiring groups of animals to accommodate their manner of social bonding.

As any naturalist can attest, particular species tend to be found in particular environments, and most species display a small, characteristic range of social groupings. While social strategies respond to immediate ecological conditions, most species are predisposed to relate to resources in certain characteristic ways and to organize themselves in ways shaped by environments experienced consistently over time.

The stability of environments deeply affects these aspects of animal adaptation. Feeding opportunities, territoriality, relationships among individuals, the composition of social groups, all are tied to the reliability of resources. Fitful change and habitat uncertainty cut against the grain of these relationships. The framework of biology deals well with how an organism harmonizes with the particular habitat where it is found. But the unpredictability of habitats over great periods of time gives further definition to a species' adaptive possibilities and continued existence.

It is apparent to everyone who has probed the natural world that organisms are closely matched to their habitats. The serrations of change in local and global environments over the period of human origin must be placed beside this profound insight, which leads to the central question: With deepening disruption of habitats, how was survival assured?

Organisms have two basic ways of accommodating to inconstancy in the conditions of survival: They either track their favored environment or they expand the range of environmental conditions in which they are able to thrive.

In the first instance, an organism evolves a more mobile approach to life. As its favorite habitat spreads, recedes, or otherwise moves across the terrain, the organism tries to follow. Tracking over large areas puts a premium on mobility. Wide dispersal offers a degree of assurance about resources. The animal can continue to associate with familiar conditions as they move, rather than confront a drastically new environmental regime. The components of a particular habitat may even be synchronized in their movements over the terrain. If aspects of the physical environment, vegetation, and animal species all migrate together, habitats may retain a certain steadfastness. Resources may stay secure in the eyes of the organism, predictable within familiar limits.

Suppose the gray squirrels of New England and the red colobus monkeys of Uganda followed their woodland acorns and forest figs to the refuge zones of their respective continents. The ability to migrate was not an adaptation to woodland or forest, acorns or figs, but a part of the movement of habitats over regional and continental scales, a response to environmental variability. The pace and degree of variability are controlled by factors far beyond the local habitats in which squirrels and colobus monkeys live today. Resource tracking is an adaptation to environmental variance, which caused habitats to be dragged back and forth across Pleistocene landscapes.

Some animals migrate over vast areas within an annual cycle. In the wildebeests of Africa, long-distance mobility is a response to seasonal shifts in climate and food supply. But for the vast majority of organisms, key resources stay put, stable for some length of time until mobilized by factors beyond the control of the organism. The extremes lie outside the experiences of a year, a lifetime, or even several generations. If they are to persist, populations of organisms must accommodate to these longer cycles of alteration. The ability to track long-term environmental fluctuation is a latent but critical tool in the survival kits of living lineages.

A second line of response, more challenging from an evolutionary point of view, results in a different kind of flexibility. If a population of organisms does not track its favored habitat, its only other defense is to accommodate to new environmental conditions by broadening its physiological limits, its diet, and other strategies of survival. The advantage goes to individuals that can buffer any shift in life conditions.

Tolerance of unfamiliar environments; the ability to switch from one favored food to another; ways of using novel resources; wider social response to diverse habitats—all these are outcomes of evolution that accent the flexibility of the organism. They are evolved tendencies that buffer environmental change and maintain survivorship and resource stability in the face of instability. Such tendencies were at a premium in Pleistocene conditions. Standard, everyday matching of an organism to its habitat was vital. But natural selection also sharpened its scythe on the entire spectrum of environmental alteration. Behavioral flexibility resolved discord from one geologic moment to the next in the conditions of existence.

Stashing away vital resources in underground tunnels or nests, as some organisms do, is one way of dealing with seasonal extremes. Storing information in the brain is a way of dealing with longer variations, enabling an organism to adjust to changes in the rules of competition and reproductive success. Manipulating the environment may be the ultimate means of satisfying needs under varying conditions: Elephants uproot trees to find water in dry times; chimpanzees use stones to crack hard nuts and make "fishing" twigs to extract termites at certain times of year.

During the Pleistocene, larger and larger fluctuations favored new ways of transmitting information and adjusting behavior and social stategies over time. The boldest examples are the technological innovations and cultural capacities of modern humans, and the highly versatile ways in which we group ourselves, defined by diverse patterns of marriage and kinship.

We have come full circle to the traits underlying the ecological dominance of human beings: The wide dispersal of *Homo;* flexibility of behavior; rapid social change; innovation in technology; and an ability to alter habitats to secure the flow of resources. This defining basis of human dominion was conceived late in the story, along with the deepening tides of nature's own disruption.

7

Complex social life, manipulation of words and meanings, the effects of an enlarged brain, cultural diversity—these fundamentals of human behavior have a life after death, echoed and shadowed in the reaches of archeological time. Let us consider how each of these emergent powers became part of the human condition.

The primate background of sociality, intellect, foraging, and manipulation is vital in grasping this process. In all these features, chimpanzees in particular display propensities often thought reserved for the human lineage. They have learned traditions of activity, passed from one generation to another. They share food in certain situations. They are capable of simple communication by combining visual symbols. Apes and monkeys are among the brainiest and most social of all mammals. The distinctive character of modern *Homo sapiens* arose as a composite of extremes and evolved under the great arc of environmental variation.

To inspect the archeological record is to look through distorted lenses. Our eyes must adjust and allow faint images to be teased from the darkness on the other side of the glass. Our first work is to uncover and reveal what no light has reached for ages. Once this is done, we must gather data, wonder and doubt, pose the most difficult questions: How do we recognize behaviors that aren't quite like our own? When did the familiar ways of the present human first arise?

Mainstream thinking on these issues has been colored by beliefs about the fixed condition of the ancient human past. The key assumption is captured by one of the most-cited remarks in anthropology over the past thirty years: "Cultural Man has been on Earth for some 2,000,000 years; for over 99 percent of this period he has lived as a hunter-gatherer."

This great icon of anthropological thinking sees all except the last 1 percent of human life as a relatively unchanging state. For nearly 2 million years, the hunter-gatherer was the primal, steadfast, natural man, living in and reflecting the status of nature. Indeed, he can still be found intact among the few tropical peoples who still practice this way of life today.

After considerable attack over the past decade, this thesis is now embraced only with marked qualification. Still, the search for the primal way of life—which, according to this same canon, requires an equally primal environment—resides in the anthropological soul. Recent attempts to reconstruct "the model environment" or "the niche" of early hominids are little more than static vignettes, so well fastened to a belief in the unvarying parameters of nature that the present is treated as if it were the past. The modern Serengeti provides us with the plan of the original "scavenging niche." A modern East African lakeside offers the model habitat where early hominids lived as toolmakers and meat eaters. And prehistoric sites spanning hundreds of thousands of years are merged together to draw a portrait of, against all evidence, the single environmental regime of human ancestry. What does the jagged edge of habitat change during the Pliocene and Pleistocene imply about such stagnant images?

Over the past several million years, a code of environmental instability was tapped out over the globe. We presume that this code was received along the human line—and other lineages of organisms—as its signal grew more powerful over the past 1 million years. In the light of this code of deepening uncertainty, let us probe the prehistoric record to see what it preserves of the foundation of the modern human lineage.

Home Bases. Fathers and mothers and children usually leave the house every day. Most members of a family lose sight of one another on a daily basis, sometimes separated by many miles. They leave the home only to return. Even if they don't return that day, it is in the sacred pact of every human society that the caretakers ensure the shelter, safety, and provisions of the young and of the spouse, the sick, and the elderly who remain at the home base.

The home base, around which a family aggregates, has long been a potent symbol in the scientific study of our origin. It is the central magnet in the human quest for food. Resources of all kinds, from food to wampum, are obtained on the outside and brought back to the family base, the center of social life, *home.* In societies worldwide, it is the place where infants are nurtured, children have the scope to play, and the elderly may be provided for. The sick may get better by resting there, and adults play out the main emotional tracks of their lives.

In the 1970s, archeologist Glynn Isaac wrote several very influential papers pointing out that in contrast to people, apes and monkeys tend to feed individually, even those that live socially. The silver-backed male gorilla and his harem feed together in a group; yet each female gleans her own handfuls of leaves; each young adult and subadult gorilla strips tasty twigs with his own fingers; and the dominant male satisfies his hunger at the food source.

In chimpanzees, the social glue is far more pliable. Individuals weave their own paths through the woodland, coalescing with and separating from friends and acquaintances. They gather around the same fig tree if it is in fruit, and they often create enough ruckus to draw others to ripened troves. But the food is rarely, if ever, taken to a different place to feed another individual. Male, female, and the immature alike journey to the sources to pick morsels for themselves. They tolerate the rare individual who begs or scrounges food, but even then the meal is consumed at the food source.

Many species of animals feed their offspring at a den or nest, and primate mothers provide for their young over the course of a foraging trip. Isaac was interested, however, in the uniqueness of human sharing. He proposed three distinctive qualities: First, food is usually removed from the source, which temporarily postpones its consumption. Second, food is regularly taken to a central point. Third, adults develop regular relationships of sharing at the home base. None of these qualities is apparent in the feeding behaviors of our closest primate kin.

Isaac, who graced his profession with a sparkling spirit and rare breadth, dug at Lake Turkana and observed the digs at Olduvai. In the unearthings of stone tools and broken animal bones, he discerned the remnants of human home-base behavior and much of its present social significance. Isaac's hypothesis, written with striking clarity, is that the social dynamics of the modern hunter-gatherer are apparent in even the earliest clusterings of exhumed remains.

Nearly two decades before, researchers studying living primates in the wild had speculated on the requisites of a home base. According to Sherwood Washburn and Irven DeVore, hominids dependent on a home base had to neutralize the predatory dangers consistently faced by other ground-living primates of the savanna. The safety of the group was essential in establishing a base camp on the ground. As later ethnographers emphasized, the highlight of the home base is its rich social matrix. This was something, DeVore and Washburn postulated, that couldn't have existed without guarding against predatory chaos in the very center of the home.

The evidence on which Isaac and others relied was found in the Olduvai Gorge. The sites excavated there by the Leakeys were thought to reveal the origin of humanity's central core—an enlarged brain, technology, culture, hunting, and the powerful syndrome of the home base. In the broken bones of animals uncovered by excavation, Isaac saw what he believed to be a distinctively human act: Hominids found meat but put off eating it until they had carted it to a central place to share with others. He postulated that plant gathering formed the baseline of the early human diet. Carrying two foods to the home base implied that the work effort was divided. Again, modern hunter-gatherers fueled the vision, suggesting a division of labor by gender: females as gatherers of the stable food sources, males engaged in the risks of finding meat.

The separation of activities by gender seemed to imply that hominid males and females had coupled up. Pair bonding evolved not simply on the basis of sexual attraction but by the sharing of food. The separate yield of female gathering and male hunting created a new kind of social currency previously unknown in the animal world. A crucial shift in male-female relationships was struck. Symbolized by a hand offering a meaty morsel and a reciprocating offer of small tubers, the home base became established. At its core was family life, one hominid nurturing another.

In retrospect, the astonishing fact is that there were no alternatives to consider. If hominids had made prehistoric concentrations of garbage, the only model that could make sense was the home base. Three decades before the excavation of Olduvai, the British anatomist Sir Solly Zuckerman had already laid out and applied the home-base model to whomever the oldest toolmakers turned out to be. Before the clues were discovered, Zuckerman had ascribed food sharing, sexual division of labor, and pair bonding to the oldest site-making hominids. The core of modern home-base life was inevitable in even the crustiest toolmaker and was considered an essence lodged deep in our ancestry.

Much of my own technical writing has focused on home bases and the scientific evidence of this keystone in the evolution of human social behavior. Since 1980, the home base has been widely debated. The details are many, but the upshot is much as I summarized earlier. Besides the telltale marks of butchery and hammerstone breakage, the bones from Olduvai bear the dental signatures of carnivores—large cats, hyenas, jackals, and smaller predators and scavengers. Both hominids and carnivores were active at these sites and fed off the spoils of collected carcass parts.

The toolmakers did not display even rudimentary acts of deterring the interests of predatory carnivores. Hunter-gatherers today usually fragment the bones of animals they bring back to camp; the grease can be boiled out of them into a soup. Domesticated carnivores (dogs) feed on the remainder. Olduvai toolmakers apparently applied no such predator prophylactics. If they required the safety of the home base, and if they distributed their food throughout the group, it is curious that they did not also use up and destroy the parts that drew potential predators to their homes. Hominids did break up bones, but carnivores had access to others rich in meat and marrow. There was much to lure the four-legged predators to the very spots where toolmakers had carried and cut up legs, heads, and other body bits.

On the basis of measures that test the idea, these early sites offered no safe refuge on the ground, no haven for sick or young, no harbor for the kind of complex social behaviors centered in the home bases of modern humans. The supposed "best cases" for the existence of such fossilized campsites, as far back as 1.8 million years, no longer provide much support for that interpretation.

What we do know is that the Olduvai toolmakers were drawn to basic features of the landscape—carcasses, plant foods, and water; trees and outcrops; and convenient places where stone was left. To account for the very early sites, all we need to posit is the recurrent visitation of these features.

When was it, then, that the home-base package first arose? At what point can we perceive, on the other side of our archeological lens, the earliest signs of a truly human campsite? All we have to go on are residues. Archeologists know that home-base activities leave behind a rough, disjointed skeleton—the bare bones of social bonding, of life centered around the home. Traces of huts or other shelters reveal the convergence to a central place on the landscape. Hearths or fireplaces suggest social cohesion, and outlines of debris defining multiple shelters or hearths hint of food shared within families.

The oldest consistent signs of home-base structures appear rather late in the prehistoric record. Hearthlike depressions of burned bones have been discovered at Vértesszöllös, Hungary, dated about 300,000 years old. In Germany, at the site of Bilzingsleben, multiple hearths and structures that appear to be shelters are known, dating to 350,000 years ago. Vague modifications at the site of Terra Amata may be post-holes indicative of structures perhaps 400,000 years old. Layers of charcoal, apparently burned by archaic humans, are preserved at Zhoukoudian, China, as long ago as 460,000 years. Although debates about the exact interpretations arise, these sites have yielded the oldest widely accepted evidence of hominid control of fire and the construction of shelters.

Burned bones from Swartkrans cave in South Africa suggest a connection between hominids and fire as early as 1.5 million years ago. We do not really know, however, how these bones entered the cave, or whether landscape fires singed the bones before they entered. At East Turkana, Kenya, burned patches of sediment are associated with hominid stone tools about 1.6 million years old; but patches and fragments of reddened silt pervade the strata at this site, hinting that savanna fires were widespread and creating doubt about whether hominids ever controlled fire as they did during the mid-Pleistocene.

Less than 200,000 years ago, the Mousterian people of Eurasia made distinctive hearths, recorded as concentrations of ash sometimes surrounded by rocks. Rings and other organized piles of bones suggest that these hominids—the Neanderthals—also built structures. Only over the past half a million years, therefore, have the signs of home-base behaviors become clear enough for archeologists to recognize, sometimes with great difficulty.

Despite changing ideas about the development of home-base activity, Isaac's hypothesis captures a basic truth. According to him, food was the central, evolutionary archway to reciprocity, the tendency to give resources or assistance to others in an often unspoken pact to be activated at some future time. Assistance may be returned by the original receiver, or by others. Human exchange is extended over networks of individuals, and it is especially strong among close kin. Reciprocity doesn't always work, but the deep-rooted concept that it *should* work is encoded in the pacts and moral visions that develop as people bind themselves together in social groups.

Whether it is sharing food or giving gifts, exchanges between people create links, establishing avenues of help *if* and *when* it is needed. Reciprocity is a way of planting seeds that may grow into the most fertile function of human sharing—the buffering of risk, uncertainty, and failure.

At some point in life, virtually all human beings undertake ventures that don't work out. While some individuals draw from within, relying on their personal resilience, an intricate, invisible web of social buffering operates in virtually all cases. We call upon our alliances. In modern societies, these range from simple friendships to complex institutions that allow a person with a fractured skull to mend in a hospital, or a man who has lost his job to exchange pieces of round metal for vegetables at a market. Such fantastic webs are awesome projections of originally simple webs of social altruism, and offer a way to buffer loss and uncertainty. If, as seems likely, food sharing nourished these propensities from the outset, the home base was the hub of such basic, enduring connections.

Isaac noted that reliability and uncertainty tend to offset each other. Reliable plant foods buffered the uncertainties of the hunt. The hunted hippo or scavenged zebra, in turn, offered a nutritional bounty far exceeding the usual collection of plants. Isaac's bold claim was that this interlocking of the food quest—the sharing that had to follow—was the most ancient, uniquely human buffer, and of powerful consequence.

The cooperative food quest requires an interdependence among foragers, who spread the risks of hard times, multiply the benefits of plenty, disperse and then meet up each day at one chosen spot. The initial benefits of the system were in the tradeoffs among individuals, the lack of a guarantee that any one individual would always be successful in the search for food. It was a system geared toward unpredictability as food supplies varied and habitats moved from one state to another. Connecting one person to another and yet another, reciprocity was the primary means of buffering environmental variability.

The first clear signals of social cohesion at a home base appeared during the great environmental excursions of the mid-Pleistocene, the interval when behaviors able to buffer the instability of resources were most heartily promoted by natural selection. The greatest periods of uncertainty occurred at irregular intervals. The power of selection was exerted in surges, varying with the intensity and direction of environmental change. The advent of the base camp was the guarantee that other foragers would meet at a given place. Reciprocity in modern human terms was born and matured during the era of mightiest ambiguity in the habitats of our ancestors.

8

Culture and Language. The anthropologist's idea of culture is rather like the strange object that keeps appearing in the film *2001: A Space Odyssey,* the thing that the hairy, long-armed man-apes reach for and touch, which endows them with human destiny. I envision the Monolith of Culture as less sleek, less smooth-faced, more of a grand old edifice, enormous in its height and weight.

The social sciences regard the concept of culture monolithically despite the plethora of definitions, which are a nightmare in any freshman's introduction to anthropology. Among them is this fairly standard rendering: "The body of socially learned beliefs, traditions, and guides for behavior shared by members of any human society." The culture concept is tailored to human beings, a practice widely adopted by anthropologists, including specialists in the study of human origin.

Culture is treated as a kind of essence, inculcating all who touch it with a particular style or form of culture. Cultures, with an *s* on the end, are the hard entities that result—separate bodies of knowledge, rules, ethics, and ways of acting. They can be identified as one culture or another: Eskimo culture, Maori culture, Japanese culture, Sioux culture, Maasai culture, and the cynic's oxymoron—American culture.

The dilemma of the Monolith is that it leaves no room for its own origin. Culture and nature are separated by a precipitous divide. What is lacking is culture's connection with the natural world. For all of its innumerable definitions, there is no currently accepted concept of culture that takes into account the origin of human beings in nature.

Prehistorians seem content to think, in a muddled way, of stone tools as "the beginning of culture." Hominids learned how to make things,

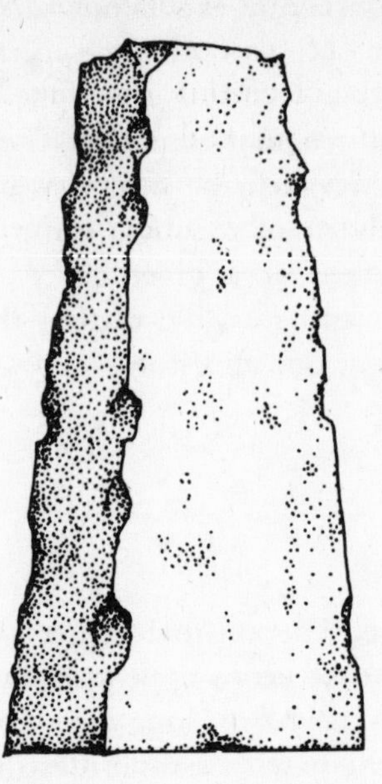

and speech was necessary in order to learn how to do it. Culture arose slowly out of human tampering with rocks, attained momentum, and finally exploded in the exponential growth of technologies and social change of the present.

In this approach, culture becomes its own explanation of human origin. It attempts to explain humanness in terms of something that is already supposed to be human. "Culture makes man makes culture," a never-ending cycle; or "culture, the maker of mankind." Mainstream anthropology has always looked to the special traits of humankind to explain the special traits of humankind. This inward-looking approach gets us nowhere in understanding human origin. Except for equating stone tools with culture, hardly a word is spoken about how the *capacities* to develop cultural behavior came about. The origin of culture is like a virgin birth, mysterious, uneasily defined.

The heart of the matter is more painful: If we find that no bridge spans the divide between culture and the natural world; if we see no comprehensible way of deriving human and artifice from animal and nature, then the human species is truly separate, unconnected, lacking real kinship with other living beings. Human culture and dominion in nature then arise from our isolation, our sense of being opposite nature. And our seemingly inevitable impulse to menace the natural world begins to make disturbing sense. There is something important at stake here.

Is there a new concept of culture that takes into account people as evolved beings, with an origin and kinship with other animals? This fundamental fact, if it were truly embraced, would crack the foundation of separation and specialness that underlies certain perceptions of our ecological dominion.

How did this thing called culture come into being? What exactly is culture? There are four ingredients involved in the answer to these questions: our understanding of other animals, the human domain, the prehistoric bridge, and the context in which the Monolith emerged.

The first ingredient is the recognition that learning, simple toolmaking, and traditions of behavior passed from one generation to the next are conspicuous in the nexus of primates where our own origin began. The stale myth that ape and monkey society, supposedly chaotic and impulsive, contrasts with the learned orderliness of human society, must wither away.

Illustrations of this point are so numerous that we need only a few here. The story of Imo, the Japanese macaque, is the most colorful. Japanese monkeys have been studied on the island of Koshima since the early 1950s. When the study began, the monkeys inhabited the forest, where viewing proved very difficult. So, in 1953, Japanese scientists introduced a new source of food, sweet potatoes, and placed them on the beach just beyond the forest, to urge the monkeys out of the trees. The macaques went to the beach and ate the sand-covered yams. What happened next is fully recounted by S. Kawamura, M. Kawai, and other Japanese researchers.

One day a sixteen-month-old monkey, a female named Imo, carried a sweet potato from the beach to a nearby stream, washed the sand off the potato, and ate it. Field-workers spied Imo doing this in September. A month later, a female playmate of Imo's followed her lead. Three months later, Imo's mother and a male in Imo's age group began to wash their potatoes. Ever so slowly, the technique caught on. First it was mainly Imo's peers who washed potatoes. Then mothers learned from their offspring.

After some time, potato washing was switched from the brook to the sea. After four and a half years, only 18 percent, or two out of eleven adults, had adopted this behavior. Yet fifteen out of nineteen younger monkeys (79 percent) washed their potatoes before eating them. Later, virtually all the monkeys had started washing potatoes, dipping the spuds into the salt water between bites.

Some years later, the researchers placed grains of wheat on the beach. The grains were almost impossible to separate from the sand, and considerable time was devoted to picking out the wheat. Imo was about four years old at the time. One day she scooped up a handful of the wheat-sand mixture, hauled herself over to a quiet pond by the sea, and threw the wheat in. As the sand settled to the bottom, Imo cupped the floating grains in her hand and quickly consumed them.

Once again, this activity was copied, initially by the juveniles and then by the mothers. The adult males were rather slow on the uptake and never really caught on to Imo's trick. Eventually they died, and the offspring of the next generation, male and female alike, adopted the behavior. At last report, the monkeys are now regular beachcombers; they swim in the ocean and have even begun cracking open shellfish, which they seem to relish. Who the inventor or inventors of this last activity were, I have never heard.

Imo and the Koshima Island monkeys dramatize the transactions of social learning and the dynamic proceedings in the lives of higher primates. They project what some primatologists consider cultural behavior. Imo's macaques varied in what they learned; they differed in their adoption of novelty and their imitation of it. Whether or not we use the term "culture," these observations illustrate the social dynamics of a behavioral tradition passed across generations.

The scientist within us must, of course, consider the alternatives: It's been suggested that each of these monkeys, adult males excepted, learned to wash food on its own, by trial and error. If so, there are two possibilities: Each macaque was an innovator equal to the one before—none of them saw Imo and all were as insightful as she (an unlikely suggestion)—or, by observing Imo and later food washers, each monkey felt impelled to try out these activities, at which point trial-and-error learning came into force. This second possibility would surely qualify as "social learning" by anyone's reckoning, even if Imo never actually showed the other monkeys what to do. Although the innovations may not have been transmitted by direct social interaction, they were adopted by virtue of experience in Imo's particular social group.

The upshot is that Imo's group adopted new behaviors which transcended a single generation. The Koshima Island macaques assumed new eating habits and ventured into new places—matters which, over time, might make a difference in the survival of the population.

In one field study after another, we are impressed by the many demonstrations of the organized nature of primate societies. A society of monkeys or apes consists of a matrix of roles and relationships learned by individuals. This matrix is found not only in patterns of interaction between individuals, but in feeding and other activities that are learned socially, including intriguing behaviors passed on across generations.

Consistent differences in eating habits, tool use, and grooming behavior are found among groups of common chimpanzees. Like all primates, chimpanzees clean the hair of other members in their social groups. In the Kajabala group in the Mahali Mountains of Tanzania, the chimps adopt an unusual posture when doing this. Partners clasp hands, raise them over their heads, and hold this posture while they groom, a technique that has never been observed in the well-studied chimps at Gombe, about one hundred miles away.

Catching termites is another example. At Gombe, the chimpanzees find a suitable twig, do not bother to peel the bark, and insert either end into a termite burrow to extract the ants. At Mount Assarick in Senegal, chimpanzees usually peel the bark from the twigs and use only one end. In Equatorial Guinea, the chimps take large sticks to pry open holes in the termite mound and then pick out the termites by hand. In other areas, where termite mounds are abundant, chimpanzees have never been seen catching termites.

Chimpanzees in different places vary in the ways they hunt, the tools they make, and the foods they prefer. Humans have no monopoly on tools, as Jane Goodall's pioneering studies at Gombe revealed. Nor do humans have an exclusive copyright on stone implements, as illustrated by west African chimps' use of nut-cracking stones. Moreover, tool use and other local preferences are transmitted socially through observation. These are traditions of behavior that involve a social memory specific to certain groups in different places. The social learning of behaviors over many generations cannot, therefore, be considered the exclusive domain of human beings.

As the body of primate observations has become impossible to ignore and primatologists gain a footing in anthropology faculties across the world, the terms *culture, proto-culture, precultural,* and *cultural activities* have entered the lexicons applied to the behavior of animals. Do these terms mean the same thing as the ethnographer's "culture"?

Passing behaviors across generations is part of the background of human behavior. The elements of culture in human beings begin with the richness of sociality, the apish potentials for developing and inheriting traditions of both simple and intricate activities. Thirty years ago, the existence of this substrate out of which human behavior evolved was barely known to zoological researchers, much less to the anthropological establishment. Now it is widely acknowledged. Recognizing it is the first step in turning the Monolith, the wall dividing human from animal, into a bridge that joins culture with its roots in the natural world.

In 1964, Louis Leakey, Phillip Tobias, and John Napier rocked the anthropological establishment by naming a new species that redefined the genus of mankind. Based on a jaw, fragments of skulls, and hand and foot bones dug from the deepest layers of the Olduvai Gorge, they proposed that a new hominid, *Homo habilis,* should be embraced as the first true human. The name means "handyman."

This first human, they reasoned, was a creator of tools of standard design. By knapping stone against stone and removing slivers piece by piece, an implement preconceived in the mind could be fashioned from a lump of rock. The oldest known lithic technology—Oldowan culture—supposedly provided evidence for such mental templates. There was a design for implements called *choppers,* another for *scrapers,* another for *polyhedrons,* and so on. Each type of tool was considered a separate target produced by adept hands, learned and taught by generations of "handymen." *Homo habilis* was considered to be this first architect, the original bearer of culture, the primal practitioner of speech and manual crafts. He was also the first hominid with a notably expanded brain. The complex of culture, linguistic skills, brain enlargement, and fine manipulation supposedly explained how Oldowan tools had been fabricated and passed on from generation to generation.

Although the canon "man the toolmaker" used to be solid ground in the study of human origin, we can no longer accept either it or its offspring, the handyman hypothesis. Over the past twenty years, Oldowan tools have been scrutinized by independent researchers to try to figure out what these fabrications of the early human hand represent. In each study, the central assumptions of the handyman hypothesis have been undermined. Here are a few key points:

> The simplest tools we might imagine are natural objects altered by repeated use, or by one main pattern of action. Chimpanzees make tools of both sorts, for example, by using a hammerstone or stripping a twig. These are also the kinds of tool found in the oldest archeological sites associated with bipedal hominids. Hammerstones and anvils exemplify the first kind of implement. Stone cores, repeatedly chipped in the process of making sharp flakes, comprise the second.
>
> Nicholas Toth's studies at East Turkana, my own measurements of Olduvai artifacts, and Thomas Wynn's analyses of Oldowan tools, all indicate that a continuum of shape, size, and edge characteristics unites the various forms. As Toth notes, the chopper, scraper, and other modified forms were incidental stopping points in the simple procedure of striking stone against stone to create sharp edges. This belies the idea that the varieties of chipped stone reflect intentional designs.
>
> While chimps and other nonhuman species do not chip stone tools (although human-induced instances in the laboratory are

> known), the repetitive products of Oldowan toolmaking suggest the same kind of action patterns and ways of social transmission that are observed in apes and monkeys. Besides the absence of discrete or planned designs, Oldowan toolmaking lacked any complex innovation requiring symbolic language or intricate social rules.

Today humankind is composed of numerous distinctions, the peaks, ridges, and valleys in the complex topography of cultures. In contrast, Oldowan tools and later handaxes attest to a vast, relatively monotonous plane of lithic skills. If we were to seek the first purposeful design of human manufacture, archeological consensus would surely nominate the handaxe, which appeared more than a million years after the oldest Oldowan tools. We can even track the struggle of hominids trying to replicate handaxes in almost any type of stone they picked up, in the awkwardly ovate, pointed shapes chipped in calcrete, sandstone, coarse quartzite, and vesicular lava—rock types that virtually defy controlled breakage, much less shaping by stone percussion. In some places, like eastern Asia, handaxes were hardly ever made, but here, the basic Oldowan-like practices of knapping stone kept on being used. This perpetual duplication contrasts dramatically with the volatile cultural proceedings of the past 40,000 years. Diversity, novelty, and cultural identity capture the imagination and attention of modern humans. Regularity and sameness occupied the minds of the handaxe makers just as rigorously.

The oldest forms of human toolmaking represent highly conservative traditions of behavior. From the conservative character of the Paleolithic, we can deduce certain fundamental properties of early human behavior, which comprise the base of the culture Monolith.

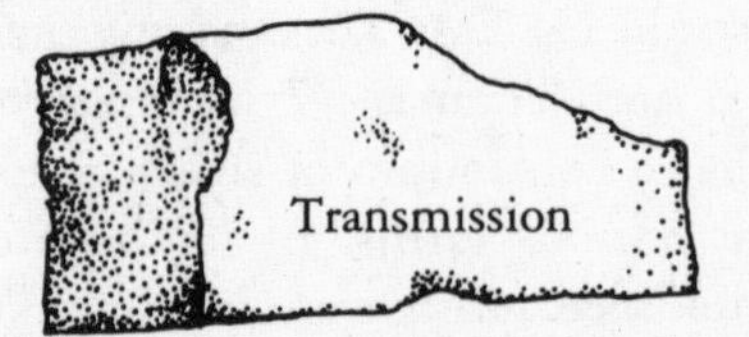

What if we had to rebuild our concept of culture from the ground up? If we were assigned the task of finding the continuities between the "culture" of the Oldowan and that of today, where would we start? Let us consider the basic operations on which the oldest lithic traditions depended.

First, toolmakers had to transmit information of some sort among themselves—behaviors or skills or ways of acting toward the environment, including other members of the social group. *Transmission* demands observation or instruction. Thus the transfer of information between individuals depends on the frequency and intensity of social encounters, and the attention paid while observing one another.

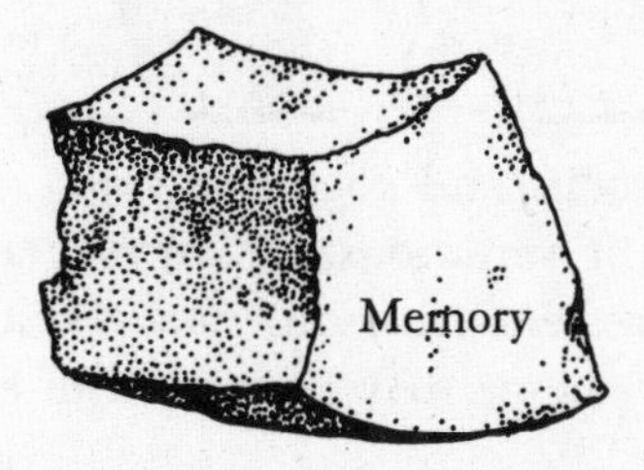

Second, social traditions require an individual to retain the information to which he is exposed. *Memory* is a good term for the capacity to store information, and the brain's role in memory determines a great deal about the amount and kind of information that can be stored. The information-storage potential may be altered over time by changes in the number of and connections between neurons.

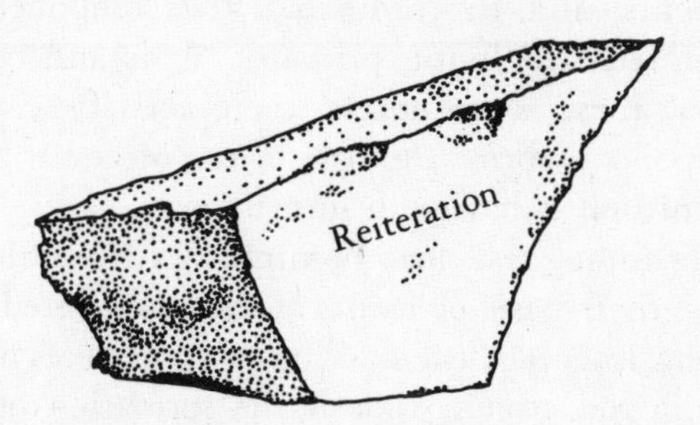

It is one thing to remember something from a social encounter. It is quite another to replicate it in the manner of Pleistocene humans. A third element in the chemistry of social tradition is *Reiteration,* the tendency to reproduce or imitate stored behavior patterns or other transmitted information. The exact cognitive causes of this tendency are unclear to me, but it seems reasonable to join together under a single rubric the various neural pathways, and the social influences on them, that enable transmitted and retained information to be translated into an animal's behavior and mental functions.

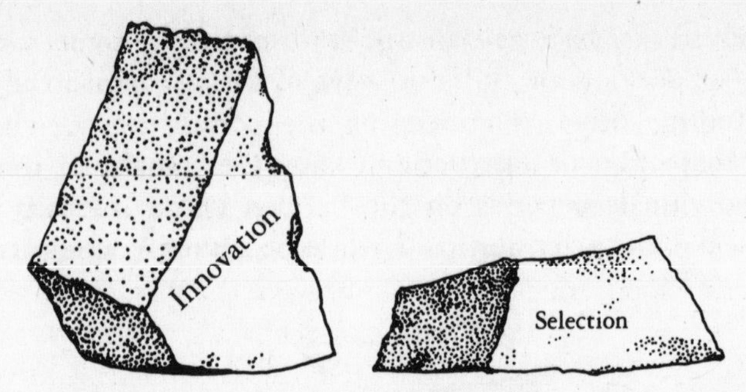

Individuals use their social and cognitive skills to solve problems, and they can manifest slight variations, which may appear to be mistakes, in imitating transmitted information. As a result, new skills and variations on a behavioral theme may occur within a group. This capacity is called *Innovation.*

Not every individual variation or type of interaction with the environment, including the social environment, is retained or reiterated. *Selection* is the term that applies to the processes by which individuals and social groups filter new information, which is then retained and applied on a selective basis.

Behavioral traditions passed across generations arise from the merger of these five elements, and the same baseline components sustain the traditions encountered in living primates. Apes and monkeys display various alloys of these elements in their activities, and interaction among the components forms the base of the Monolith, the foundation that underlies cultural behavior in humans.

There is little to suggest that hominids in the oldest periods had access to any but these basic elements of socially learned behavior. Over astoundingly long intervals, the making of stone tools and the creating of clusters of chipped stones, indeed the formation of the entire archeological record, are all consistent with the passing on of simple behavioral traditions. These do not, however, reflect the monolithic entirety of culture.

The stone products of the Oldowan and Acheulean toolmakers reflect a low degree of innovation. Variety—the creation of new tool styles and methods of manufacture—was so constrained over such long periods that the mental and social processes favoring reiteration must have been ruthless in dousing the spark of innovation.

Duplication of the handaxe seems to have had parallels in the lives of other primates, however. If we were transported a million years back in time, no one would be surprised to see an ancestral chimpanzee practicing ways of getting food essentially as he conducts them today. Although certain aspects of chimp tool use may have evolved in response to Pleistocene fluctuations, it would not shock us to see chimpanzees cracking nuts with the same kind of stone hammers that some groups now use. Or we might see them using twigs to entice termites into becoming food, much as we can observe them in the African woodlands today. There are many such instances where specific learned activities have probably been repeated over enormous lengths of time.

In the archives of the past 2 million years, we can discern flashes of hominid variation and imagine situations that invigorated the transmission of certain activities as hominids came together around sources of food or stone. Turning points undoubtedly occurred in the battle between innovation and reiteration. Shifts took place in the tension between social memory and forces like population decrease, or the disappearance of a resource, which tended to dissolve social groups or the information they retained. That the behavioral traditions of early hominids and living apes arise out of the same basic components of social learning does not imply that chimpanzees can replicate Oldowan toolmaking. The manual and neural skills needed to make a chopper or a handaxe are not the issue; our aim is to discern the capacities needed to pass any kind of skill or other information across generations. Transmission, memory, reiteration, innovation, and selection are one way to describe the processes that account for behavioral traditions.

The Monolith of Culture is still incomplete. Let us discover its other building blocks.

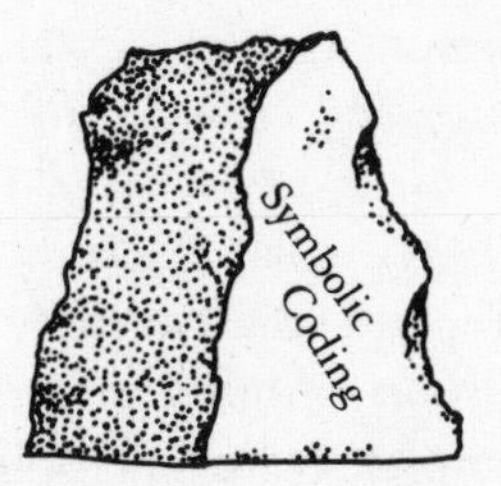

Every day we see, hear, and produce special codes that supply information. Combinations of sounds and signs are spewed forth, and somehow we tend to understand each other. This is *symbolic coding,*

translating information through creative weddings between things and meanings.

Symbolic coding enables us to construct large edifices of social and personal significance out of the most inconsequential building blocks. We assign meaning and significance to sounds, images, objects, and almost everything else in the environment. By stringing these symbols together, we create languages and rituals and special social memberships.

The result is the drenching of human life with symbols, which saturate both our private thoughts and our public personae. This uniquely human way of labeling encompasses all the symbolic devices—language being a primary one—that influence behavior and extend our social definitions and expectations. These definitions are seen especially in the handles of kinship, status, and role that we apply to one another. Even when we die, the symbols and social definitions connected with us may carry on; they have a continuity of their own.

There are specific mechanisms in the human brain for filtering sounds, assigning meanings, generating combinations, building and interpreting grammatical rules. Regardless of society or place, human infants follow a universal pattern of learning a language. As children, we all went through a standard series of stages, babbling our way in an intricate grammatical forest, trying out the permissible routes in it. Our brains were primed to learn the combinations of trees we could cut in that forest, and, ultimately, we put the linguistic planks together in ways that made sense to the leviathans that ruled around us.

All the possibilities of human language are founded on the internal circuitry and programs of the brain. These software programs are open, waiting to suck in and process information about the world. Environmental inputs are necessary to engage the adaptive possibilities of those programs. Symbolic coding thus entails the construction of an internal reality about our interaction with the external world.

Why did this capacity for using symbols and creating symbolic worlds evolve?

Chimpanzees and gorillas display the rudiments. They learn meanings, manipulate symbols, and put a small number of them together. Symbolic coding depends, however, on a highly magnified array of mental and social possibilities. The complexity, inventiveness, and versatility of spoken and signed language are ample testimony to a broad potential.

A plethora of theories have offered reasons why language evolved in humans. Most of them emphasize the fact that language is a form of social communication. Perhaps language evolved because of the value of stories in relaying information within the social group, or because it reinforced the bonds within larger social groups, or as a clever way to deceive a competitor. And so on.

In our attempt to grasp the context of human origin, is there anything about symbolic coding that indicates the environmental milieu in which it developed? I believe there are two such aspects, both consistent with our thesis.

In everything we say and much of what we mean, there is a certain proclivity for variation. Although infants in all cultures display a standard pattern of acquiring speech, each also gains access to a system of new utterances and meanings, errors, things never heard before. The history of language—the changes and divisions of utterances and their meanings over time—is well documented by linguists. We say that languages evolve. It is the mutable, low-fidelity, slip-of-the-tongue character of the human mind's connection with the language centers of the brain that provides the basis for this phenomenon. To all appearances, language is poorly designed for repetition, though repetition may serve it quite well in passing on oral traditions and verse.

The innovative property of language may have evolved in relation to social contingencies, and I suspect that these contingencies were most volatile as the environmental resources underwriting social strategies underwent extreme fluctuation.

Symbolic coding also enables people to refer to the invisible, allowing information to be passed on in ways that transcend the need for direct observation. Symbols may be created for things not immediately seen: a past event stored in memory, a river many miles away. A simple icon or a crude likeness may stand for something that elicits intense sympathy or emotion. And symbols may be created for things never seen: a principle, an abstraction, an aspiration, a god.

We say many things that could easily be represented by mere signals or acts. Imagine a system of signals understood and accepted by everyone in a social group. With different tones or gestures, people would be able to convey feelings and relationships between one another, to deliver messages about an immediate situation such as danger or peace, and to offer recollections of common experience. But truly invisible phenomena are lost without complex symbolic coding.

Consider this statement: "My grandfather told me there was a great drought when he was young; and when the water dried up in his homeland, he and his family quenched their thirst on some brown fruits from the opposite side of that mountain over there, where we have never been." This illustrates in a very simple way a basic service of all modern human languages: a nearly infinite capacity to express contingencies of time and place. Nonhuman primates are capable of transmitting information across generations, but not the invisible, so far as we know, nor complex explanations of what one would do if some other event were to happen. By all indications, early human toolmakers were dedicated to passing on knowledge and skills across many generations. But we are short on evidence that they possessed the infinite variety of transmissions inherent in modern human language.

Under what conditions did this capacity become magnified and ultimately soldered to the core of human symbolic coding? What settings favored an awareness and a sharing of the galaxy of invisible events and possibilities outside of one's own direct experience?

It has long been presumed that language, perhaps the most amazing and complex of all the higher faculties in humans, arose in increasingly open, arid, glacial environments during the Pleistocene. We now know that these were bracketed by incursions of warmer, moister, more vegetated climes, that the sequence was periodically disrupted by unpredictable events, and that this kind of variability was by no means unique to northern settings. Knowing this environmental past, we can conceive how changing contingencies—repeated shifts in resources, periodic redefinitions of the environment—furnished the vital spark to symbolic coding.

The social and mental basis of symbolic coding increased as, over time, ecological settings varied. Environments changed, local resources appeared, moved, and disappeared. The pillars of survival and advantage in one era turned into *invisible* things that really mattered in another. This kind of ecological instability reached its pinnacle within the past 700,000 years. The thick symbolic forests inhabited by the Pleistocene survivor among the hominids now begin to make sense; over the course of human existence, they buffered the cold winds of uncertainty.

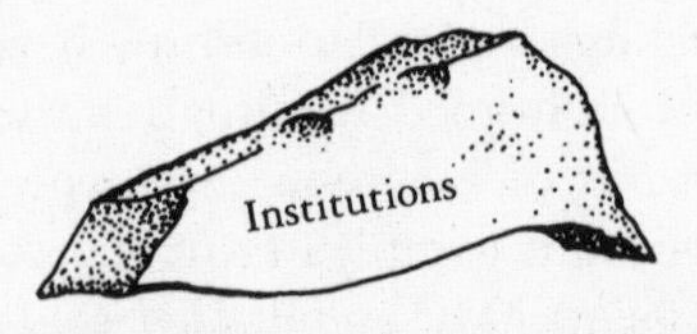

Imagine a seventh component that raises the concept of culture to its full monolithic height: *Institutions* is the name we give to the highest order of complexity in human lives. It denotes the hyperreaction between symbolic codes and social behavior. Because of this interplay, social relationships are catalyzed in a manner distinctive to human beings. The result is what anthropologists call cultural institutions, which draw together all the ways in which information and symbolic codes are organized into complex social order.

Symbolic systems build, and as they continue to interact with social relationships, there is an exaggerated "return effect." Interactions among people intensify and become charged with social meaning. People create centers of thought, creed, accepted action and specific bodies of social relationships and values called institutions, which embody most of what social scientists call culture.

Anthropologists lump institutions into special categories—kinship, politics, economics, morals, law, art. These are the ways in which people identify and express themselves socially—by their definition of family, religion, specific political and economic activities, legal expectations and actions, and socially guided routes of creativity. Cultural institutions permeate and strengthen the human world. They are the fortresses of our ordinary lives, responsible for the extraordinary power of human activities from local clubs to global alliances.

People form institutions to govern themselves and create accepted formulas of marriage and kinship. They devise schools, businesses, churches, constitutions, political parties, libraries, organizations of every kind. These are the intangible webs woven out of shared meanings and values that bind people together, and they are strengthened by insignias, songs, or other symbols of affiliation.

Institutions obviously depend on symbolic coding, much as symbolic coding leans on transmission, memory, and other older parts of the Monolith. In modern human life, symbols and meanings exist on their own, but in most cases they have an unmistakable bond with particular institutions—the colors of a flag, the music of a nation, the emblem of a favorite sports team. All of these have social connotations that ramify, build, and organize societies of people and divide them into segments. These are symbols in the service of institutions.

Because of institutions, almost everything around us has some social affiliation, a place within a truly formidable crossfire of membership and categories. In many cases, institutions carry with them such an intricate creed of social meaning that it makes symbolic coding and the basic exercises of human language appear simple and rudimentary.

By clothing or by action, by the way we speak or even the things we eat, we wear badges that define our alliances. We inject meaning into objects, making them susceptible to as many varied expressions as there are societies, languages, and ways of living in this multifarious human world.

Because we use symbols to create institutions, even the basic appearances of people—both outward evidence of race and cosmetic alteration—are manipulated as significant badges. We even take the most practical aspects of our animal heritage—gender is a primary example—and inflict upon them the peculiar bents and measures of our upbringings.

Wherever people aggregate, our symbolic refrains send signals, correct or mistaken, intentional or unintentional, about who we are, defined as much by others as by ourselves. These align us with one meaning and purpose and prohibit us access to other rooms within our sophisticated social edifices. Accepted, rejected; friend, foe; this, not that; we cannot help but be interpreted in some deeply symbolic and social manner.

Institutions direct individuals toward certain meanings and values. At the same time, they take on a life of their own and often persist only for the sake of their own survival. They may outlast the symbols and meanings that moved people to act together in the first place. Because people rely so heavily on them, institutions are more than just their associated symbols or meanings. For this reason, the formation of institutions is a distinctive part of the cultural Monolith. The electrified arc between symbols and sociality deserves its own distinctive label, for it radiates its own force, tempered in the furnace that anthropologists call culture.

The overlay of institutions was born when the simple symbol systems hominids carted around with them began to crystallize into complicated forms. When institutions became an organizing power in human life, communal existence began to acquire a strength and intensity out of all proportion to the social impulses and traditions we share with our primate cousins.

The tension between individual variability and the network of social beliefs and rites reached a new level of suspense. Even when innovations occurred, the institutional foundation was usually passed on intact. That is why any given culture has a lifetime, a measurable continuity of its particular ways of living. The inextricable fusion of this final element with the underlying components of the Monolith means that institutions are often transmitted as a whole; people reiterate institutionalized patterns of meaning and action in their entirety, selecting certain patterns of social action and rejecting others. This arises from the merging of institutions with the much older capacity whereby animals select from an array of possible social behaviors.

By this astonishing merger, people in groups come to share certain meanings and ways of life different from those of other groups. We create identities for ourselves by donning a series of shared social badges—and equally by rejecting the social identities of other groups of people. The culmination is the division of humankind into cultures, the amazingly diverse packets of human life, each of which has a life of its own because of the institutions that comprise it.

When anthropologists talk about a particular culture, they usually mean its institutions. Yet institutions are found only at the awesome peak of the Monolith. By breaking it down, we start to see that institutions are a vast overlay. In studying this overlay, anthropologists seldom realize that the underpinnings of cultural behavior actually evolved much earlier—and not just in human beings.

The study of culture focuses on the most complex forms of symbolic beliefs and social behaviors. The individual is subsumed, interviewed to discover the cultural institutions that surround him. Yet without knowing the multiple parts of the Monolith, there is little hope of reconciling the individual as the carrier of cultural information and the institution as the determiner of beliefs and behaviors. Dissolving the Monolith, recognizing its origin, permits us to see the interplay between the two, the relationship between personal patterns of behavior and the powerful precepts and conventions of social conduct.

The point most pertinent to our investigation is that an evolutionary bridge exists between the human and animal realms of behavior, a bridge that has been conveyed through time. Culture represents continuity rather than the separation of human beings from the natural world. Its existence rests on an old primate heritage. As human beings evolved, the unique elements of our cultural life eventually coalesced and gave shape to that extraordinary structure we have called the Monolith.

The overarching feature of the Monolith, the creation of institutions, has two basic implications for human life. First, individuals are protected in the multiwalled fortress of cultural institutions. This fortress is the ultimate buffer, the extension of reciprocity and insurance into virtually every aspect of human social life. Governments, currencies, creeds, social and economic values, expectations at every level of social functioning, all tend to buffer disruptions inside and outside the social

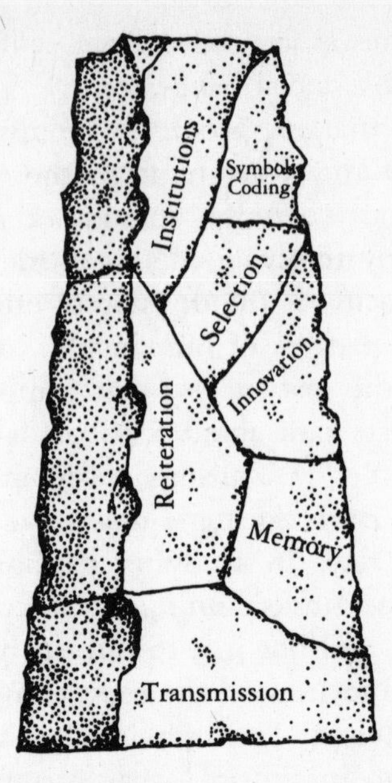

sphere. Institutions are the fundamental ways in which human beings meet the contingencies of their environment, both ecological and social.

The second implication involves the versatility of human life. Human beings develop multiple approaches to life, differences both large scale, which we call cultures, and small scale, which are the distinctions among the people within those cultures. Such variations testify to the flexibility of the species. To a great extent, diverse cultures and groupings of society are accommodations to the immense range of global environments. To another extent, the long and short division of humanity arises merely because it is possible, a probing of the human way of life.

Cultures are specialized manifestations of what it means to be a human being. They represent narrow reflections of all the possible ways in which people can get on in the world. Human beings are a collective of forest specialists, desert specialists, urban specialists, nomadic specialists, agrarian specialists, industrial specialists, small-town specialists, and globetrotter specialists. Each individual establishes his or her culture according to the surrounding possibilities, developing, editing, nurturing, cutting, maturing these possibilities. We invent ourselves out of the meanings and behavioral options around us. Each of us is a construction worker, using sets of meanings to build the specific social properties we call culture.

Because selection is inherent to the process, we all end up as specialists in one form or another. But the compiled result is a manifestation of the ultimate generalist, adaptable and responsive to contingencies. A person may not be very adaptable as an adult, but at birth almost anyone can become one or another of the possible specialists. Matching the individual and the social group to specific environmental circumstances is done most effectively through the mechanisms we call culture.

Dividing the human endeavor into divergent ways of life raised the chances of perpetuation along a rough and unreliable path of origin. I believe that the immense variability of our species and its cultural institutions was not a response to environmental stability, certainty, or steady trend. The mutable side of culture, the divisiveness of cultural variability, and the tremendous buffering power of institutions would have little reason to evolve under such conditions.

According to popular conception and textbooks alike, culture is the way human beings perfect their adaptation to the environment. But in light of the rich record of Pleistocene oscillation and disruption, perfection was hardly the tissue. The cultural capacities unique to human beings evolved not for perfection's sake but in response to changing contingencies. Having touched the Monolith, and now embodying it, human beings are a cultural organism geared toward buffering environmental uncertainty. In this lies our capacity for control and the ecological dominion of our species.

To dissect anthropology's concept of culture is an enormous task. The sheer height and breadth of the idea has always proved too difficult to scale. But the individual blocks of the Monolith are easier to consider. Cultural behavior is a magnificent alloy of social, cognitive, and behavioral faculties that have emerged over time. These elements represent a set of biologically evolved potentials whose tightly bonded federation has created a distinctive nongenetic mode of inheritance across generations. Numerous animals possess the basal blocks of the Monolith. The uniquely human part of the edifice allows an astounding power to imagine and to invent, to revise and to reject, to solidify and to diversify the social approaches of our kind.

When were the symbolic, institutional, and diversifying aspects of culture first expressed on a regular basis? The oldest lithic technologies had no inherent progress built into them. Beautifully symmetrical, finely chipped handaxes first appear in the archeological record after 500,000 years ago. But tool kits this young also contained handaxes as crude as the oldest ones made a million years earlier.

Beginning roughly 250,000 years ago, a small number of new, distinctive tool shapes were being knapped in some parts of the world, the result of a new toolmaking method in which standardized slivers were removed from a carefully prepared stone. By sometime after 130,000 years ago, a number of other new artifact forms had entered the tool kits in some parts of the world. Around this same time unique types of artifacts began to characterize specific geographic locations.

Even in this later period, fine styles of flaking stone are sometimes found to be older than rougher styles. Whatever behaviors and factors were responsible for this, progress was clearly undermined. It was only when symbols and institutions began to exaggerate the chances of innovation that some semblance of direction and progress was noticeable, at first regionally, ultimately on a global scale. Toolmaking hitched a ride, and the saga of technology has offered a shocking display ever since.

In seeking our unique cultural roots, we must discover when complex social variation was first manifested, when hominids began to perpetuate and alter their behaviors by using rich symbolic codes, and when mental operations encouraging novelty and social mechanisms encouraging diversity first came into existence. We have, however, only the most shadowy archeological clues to go on.

No convincing symbolic object older than a few hundred thousand years is known. The site of Berekhat Ram, Israel, has yielded a rough depiction of a female form on an exotic pebble perhaps 230,000 years old. As far as I'm aware, this is the oldest widely accepted object of human aesthetic expression.

As a symbol of symbolic behavior, it stands alone until after 130,000 years ago. A slight rise in symbolic behaviors was registered at about that time. On rare occasions, people like the Neanderthals engraved, notched, and carved the bones and teeth of animals. Sometimes they dug shallow pits to dispose of their dead, and sometimes the bodies were placed in a flexed position.

The modified objects and graves of the Paleolithic fan the fire of archeologists' debating passions. Strenuous arguments have tried to decide whether or not the Neanderthals possessed the *capacities* of modern humans, but we have no direct access to capacities. There is no reason to think that stone flaking, bone marking and grave digging mirror the "potential" of a Neanderthal with any faithfulness. All we know is whether symbolic coding and diversely organized bodies of symbols permeated the sphere of preservable things these humans left behind.

Nothing suggests that the Neanderthals ever engaged in the diverse, symbol-drenched systems of social action that mark the character of all human beings today. The Neanderthals lived between 200,000 and 30,000 years ago in Europe and western Asia. They could hunt and scavenge food; they made simple, decent hearths and shelters; they did many of the same basic things modern-looking cave people did later on. But for the most part, they were largely unmoved by the forces of the Monolith that assure cultural diversity across our species and make change nearly unavoidable over long spans of time.

The early symbolic displays of Neanderthals and slightly earlier humans give momentum to our study nevertheless, supporting our thesis that the behavior and ecological character of human beings matured in response to wide environmental fluctuations. A strong objection to this view could be leveled if it turned out that flexibility and environmental buffering were first manifested under relatively stable conditions, or if the key agents of human culture—words, symbols, institutions, and diversity—were divorced from the oscillations of highest amplitude over the time range of human origin. But we find that the oldest proxies of symbolic coding were conveyed relatively late in the Pleistocene, coinciding with spasms of habitat change. Looking at the entire course of human origin, the link with intense environmental variability could hardly have been stronger.

An abrupt climatic swing from glacial to interglacial conditions is recorded in the deep ocean cores around 130,000 years ago. Between 115,000 and 72,000 years ago, irregular reversals between forested and open vegetation can be seen in the records of European fossil pollen. The long return to a glacial terrain was broken by several shifts to warm-loving vegetation. Climatic swings were sometimes very fast, and environmental stability was expressed in spans of a few thousand years at most. From the perspective of a lifetime—yours, mine, or a Neanderthal's—that is a very long time. But not in the perspective of a lineage evolving over time. Under these intense fluctuations, hominids displayed new ways of wielding information, methods that would prove surprisingly effective in dealing with unexpected variations in key resources that required people to call upon things not immediately visible.

After 40,000 years ago, we see the first clear florescences of the entire Monolith. Symbols began to pervade manufacturing and burial. People decorated themselves with pendants, necklaces, and beads. Magnificent bursts of painted and sculpted symbols were projected onto stone, dark caverns, and lumps of clay. Behaviors and artifact styles diversified from one region to another. Tools of antler, tusk, bone, and stone were produced to precise specifications. The forms were often dictated by strict rules, not mere reiterations of a broad, continuous range. People began using space in more complex ways, beyond the distinction of pit, hearth, and shelter. With thousands of mammoth bones, they committed themselves to laborious feats of architecture. Political status, economic networks, and long-range alliances appear to have developed. Impaled human figures were represented on cave walls. People made bone flutes and ivory rattles; they must have invented rhythms. With exchanges of amber and trading of flints, they developed a heightened sense of desire. They wanted things far beyond their borders. In art, adornment, music, and economic production, they created nearly every kind of badge of personal and social identity that it is possible to imprint on the fossil record. As they crawled on hands and knees through tiny passages leading to vast painted chambers, they invented mystery and inscribed a deeper code of belief than any known to us from earlier times. Beliefs, aesthetics, music, murder, labor—all may have reached farther back in time. But in the late Pleistocene, tangible symbols were for the first time minted on each of these endeavors.

The developments did not appear in one full blast. Beadwork and engravings began to blossom about 40,000 years ago in Africa and Europe. The most spectacular florescence of symbolic behaviors can be traced to around 18,000 years ago, with the development of artistic forms and cultural diversity on several continents. These happenings, not coincident with any particular biological change, are our first sign of the peculiar responsiveness of modern human societies to extreme environmental conditions.

At the 18,000-year mark, habitats in the tropics deviated toward extreme aridity, while higher latitudes were encased in glacial ice. According to Olga Soffer, Meg Conkey, Lawrence Straus, and others, as hominid populations were squeezed into a smaller geographic range, interaction between them began to intensity. As a result, people began to erect walls of social distinction between themselves and other groups. If this was so, the explosive expression of human cultures coincided with the new social conditions brought on by intense climatic alteration. The key point is not the exact timing of these developments, but the fact that these fruits had ripened in the passing of several hundred thousand years of the largest natural environmental fluctuations ever registered on the planet.

If the settings of human origin had been stable, secure, and balanced, the innovating, diversifying, buffering, and exploiting aspects of human cultural behavior would make little sense as a response to natural conditions. We would have to say that culture arose as a progressive means of "improving" Earth for our exclusive benefit, and that cultural behaviors evolved to exploit the stability of natural habitats. And this is, indeed, how we have largely portrayed our relationship with the environment.

Now we are pressed to adopt a different point of view. Cultural behaviors, in the modern human sense, evolved in the midst of nature's grandest alterations. The range of oscillation during the past 1 million years was greater than in any part of the first 4 million years of human evolution. Neither stability nor directional change can, therefore, be credited for these defining elements of human behavior.

Symbolism and institutions form the towering mass of the Monolith. Languages respond powerfully and swiftly to contingency. The human species' tremendous diversity resides in its institutions. I believe that these facets of cultural behavior were honed by their role in buffering fluctuations in the conditions of survival. Their full expression late in the Pleistocene, especially over the last major cycle of global change, is, I believe, the culmination of a selective process that reached its greatest strength over the past 700,000 years. So we may offer an addendum to our definition of the Monolith. It is the great fount of human flexibility. Its evolved potency resides in the power of accommodation, the power to channel human behavior, and the power to alter the substance and structure of our surroundings.

9

Brains. The higher functions of the brain—thinking, remembering, imagining—are central to the study of human ancestry, so it is important to discover whether the evolution of these particular cranial activities provides any clues for or against our thesis. The weirdest irony in the pursuit of human origin is that we unleash our unwieldy brains on the subject of *why* these organs evolved to such a great size in the first place. Scientists and novelists alike have pondered the possible causes, and the answers are a list of celebrations and misgivings about life in general.

The brain, according to one tradition in Western thought, is connected to noble causes. The organ of intelligence, problem solving, and humanitarian enterprises, it is proof of our separation from other animals. To other authorities, it is an enormous social calculator, capable of lying, cheating, and cunning manipulation, the mastermind in the self-centered game of social chess, as some have called it.

There is no end to the suggestions and theories about brain evolution. Each underlines one or another feat of brain function: the capacity to create fabrics of thought by using language; to create things—tools, shelters, and architectural wonders; to make war in order to destroy the things other people create; to cooperate in order to share and give succor, even to those we don't know; to compete and create disadvantages for others.

The state of the debate is clear: Our brains have won; they have thought of more good reasons to explain their oversized existence than anthropologists can possibly handle. Despite the conflicting opinions, some basic points of agreement bear on our thesis.

The brain does its work—processes information, creates thoughts, makes and weaves together sensations and actions—by chemical messages between nerve cells, propagating electrical signals between clumps of nerve tissue. Each nerve cell has about 10,000 channels of communication with other neurons. These tiny gaps, called synapses, are bridged by chemicals released into the gaps by nerve cells. Multiply these synapses by the 100 billion neurons in the human brain, and the product is an astronomical number of circuits in each person's head that do things like create images, abstractions, and symbols—what we call thinking—and deal with a constant barrage of sensations, perceptions, movements, and stories, even while we're asleep. This cacophony of signals is put in order by the rate at which nerve cells ignite, the complex circuits thereby called into service, and the interaction between the circuits. The huge variety of connections in this natural computer of ours raises the level of sorting, integrating, and producing new data to a plane far beyond that of any machine we have ever built.

A human brain is not, however, an unstructured mass of neurons, grouped and connected merely by life experiences. It has a structure like any other anatomical organ, a precise layout and a design to its mental operations. (If only there were a manual and scheduled maintenance!)

There are two points of view about how the higher mental functions of the human brain actually operate. The first, the theory of *mass action,* contends that prodigious memory and data synthesis result from the total amount of nerve activity. The more nerve cells or connections, the more complex the possible behaviors and mental operations. This implies that any increase in the size of the neocortex would, over time, improve mental function. The second hypothesis, *localized function,* holds that higher mental operations are localized and stem from the activity of specific brain areas, not from the simple additive effect of neurons.

The second view has proved correct. While large networks of neurons are involved in processing sensory data, memory, and images, the most complex functions underlying human behavior are carried out in specific places—between certain furrows, bumps, and other landmarks that can be identified on a map of the brain. As one specialist put it: "Apparently unified mental life is actually composed of multiple, distinct components. Different brain areas perform different mental functions, and disconnection of areas prevents normally integrated mind function. Mentality literally disintegrates."

Our brains have centers of memory and language and hubs for integrating information from the different senses. Images, thoughts, and the links between experiences that we call learning all seem to have their special residences in the brain. The pathways and programs of the brain are data-sensitive magnets of outside information, craving it in order to do their work. The specific anatomical design of the modern human brain deals with an extraordinary range of possibilities.

Perhaps most significant of all, the brain works in the midst of other brains. Our mental organs operate in a social context, and human action and thinking are collective and interactive. The results far transcend the biological foundation and product of any one person's brain.

It is imperative that the brain operate socially, nurtured by others, because it is a very costly item to grow and maintain. It makes up only 2 percent of adult body weight, but it consumes around 20 percent of the body's energy. As organs go, that's not a very fair tradeoff, and it gets worse because our brains require such a long period of time to mature.

A newborn human is among the most incompetent mammals on Earth, largely because, at birth, the brain, in fact the entire nervous system, acts as if it were still in the womb. A baby at birth has only about 25 percent of its final brain size, about half the maturity of a newborn chimpanzee's brain. And for at least the first year, our brains keep maturing at the fetal growth rate, finally slowing to the rate chimpanzee babies assume at birth. At five years, a human child has 95 percent of adult brain size; a chimpanzee attains this mark in less than two. In the end, the chimp brain averages about 375 cubic centimeters in size, the human brain more than three times as much. So the tremendous cost of growing a human brain is not just in metabolic energy, but also in the amount of time required. Human life starts out as a precarious venture because we stay awkwardly immature for such a long span compared with other mammals. The enormous energies of feeding and caring that parents expend are, needless to say, absolutely necessary.

Our central nervous systems enable and encourage behavioral versatility. But it is not size alone that creates the amazing array of possibilities in human life. The central nervous system's maturation is prolonged in proportion to the brain's great volume. The neurons of our neocortex retain a youthful adaptability. Nerve cells make pacts with one another, and circuits are laid down for a significant portion of life. This prolonged development of brain pathways, which neuroscientists call "neuronal plasticity," is responsible for our mental flexibility. Human beings are born into a world full of options and risks. It is the brain's lengthy, out-of-womb maturation that underwrites our vast abilities to pick up data about the world, and also our susceptibility to mental trauma, deeply entrenched in the early years.

Woven into the whole process is the social context of childhood. Human adaptability, including its potentially devastating risks, cannot be uncoupled from social nurturing. It is impossible to disconnect human flexibility from the intense crisscross of social care and conflict.

The tremendous obstacles posed by helpless immaturity are the reasons that a brain like ours is a most unlikely structure to have evolved. And so it is generally assumed that the brain's tremendous increase in size and complexity required intense competition or inordinate tests of survival.

An important theory of mental evolution fueled by this perspective is the theory of social intelligence. In 1976, zoologist N. K. Humphrey wrote an influential paper, "The Social Function of Intellect," suggesting that the practical sides of life are insufficient to account for the complex functions of the primate brain. The problems posed by the external environment are simply not complicated enough.

After watching gorillas in the wild for some months, Humphrey could detect very few challenges that they faced in procuring food or otherwise manipulating their forested habitat. Assuming that this habitat never changed, he began to consider the ecological settings of gorillas and other primates totally inadequate to explain these animals' impressive problem-solving abilities.

High intelligence, he said, must have come from some other factor. Like Alison Jolly earlier and many others later, he argued that the driving force behind mental evolution is the complexity of social life. High intelligence means that organisms adapt their strategies to continually changing conditions, as in the social realm, where the level of planning and accommodating is, in Humphrey's words, "unparalleled in any other sphere of living."

In recent years, Humphrey's idea has been adopted by psychologists and biologists who claim that "much of human intelligence is social intelligence, the product of selection for success in social competition." According to this view, social environments, in contrast to natural ones, are extremely complex and ever-changing. Social life involves complicated networks of competing individuals, who seek advantage by an intricate process of alliance and trickery. The more complex the society, the more creative and calculating mental functions must become.

Another name often given to this idea is "Machiavellian intelligence." It holds that the higher mental functions of human beings evolved to negotiate social interactions, create alliances, and manipulate friends and foes. Human intelligence is played out in multiple arenas—in finding mates, influencing the behavior of others, and predicting the actions of rivals, and human mental life has been spurred by these constant social demands.

The problem with this hypothesis is not the dynamic link it supposes between mental function and social complexity, but its assumption about the natural world. The idea that human intellect is predominantly social arose from Humphrey's original assertion about the stability of environments. After an organism adapts, its habitat is unable to provide the kinds of demands that prove critical to the competitive success of individuals.

Belief in the stability of natural habitats over evolutionary time is, of course, unacceptable. Whether the environment challenges an organism or not is a matter of the time scale in which habitats are viewed. While individual organisms do, in fact, view environments over only limited periods, the lineages of which they are a part persist over a geologic time frame. This requires them to face the long-term variability of habitats, indifferent to the specific needs of organisms.

As we observed earlier, both social groups and the social strategies of individuals are intimately linked to the quantity and distribution of resources. Environmental change means that social strategies are bound to contingency. The "social function of intellect" gains momentum from the ecological drama in which a lineage evolves. The social side cannot be divorced from the ecological, and in humans, it cannot be divorced from the environmental odyssey we have sketched.

If the evolutionary character of human beings was incited largely by the Pleistocene's environmental moods, what do we expect to discover about the human brain? The special functions of the neocortex would solve the problems of diversity, novelty, and complexity. Human beings would be possessed of many different kinds of intelligences. If this view is correct, perhaps the best definition of intelligence is not to have a preconceived notion of what is intelligent. In an uncertain world, problems of survival would be handled by myriad paths of skill and insight.

The distinctive operations of the human brain are designed largely as open programs; the brain must feed on environmental inputs in order to release the adaptive possibilities of these programs. Diverse inputs and flexible outputs require certain arrangements of hardware and specific sets of programmed instructions, as anyone familiar with computers knows. This relationship between software and hardware was encouraged by an unprecedented increase in the sheer number of processing components and their organization into special localized operations.

As brain enlargement took place over the past 2 million years in the lineage leading to *Homo sapiens,* the gross count of neurons increased. The multiplication of synapses and connecting fibers was even more dramatic. The interface between specialized segments of the cerebrum became wider and wider. As a result, the brain's information-processing capacity expanded along with its basic higher functions—mediating incoming and outgoing information and memory, and taking experience into account.

The organism in which this occurred became increasingly committed to solving, absorbing, and controlling unprecedented information about the conditions affecting its existence. The *databases* of the human brain relate to specific conditions, particular habitats, and experiences. But its *programs* mirror the variability of those environments, the vast range of conditions in which human beings may find themselves. Thus we possess the capacity to absorb and to manipulate data about an enormous range of situations.

The fact that brain functions are localized also affects what we think about the origin of human beings. Language, culture, and intelligence are not simply the products of enlarging the brain and the number of neurons. They derive from discrete areas of the brain and distinct avenues of communication between the parts. That they do not arise automatically from pure brain mass means that the origin of language, abstract reasoning, imagination, social cunning, intricate strategies, economic planning, and other things that we believe our large brains are good for did not necessarily commence with the first main increase in brain volume, which is linked to the origin of *Homo,* 2 to 2.5 million years ago. One ounce of brain added over time was not necessarily equal to every other ounce. The important thing was the organization of those increments into control towers that responded to the matters of Pliocene and Pleistocene survival.

At this point in our study, it should be apparent that the higher mental operations coloring human ancestry were tightly tethered to the spatial maps and movements of resources by toolmakers, to home-base behaviors, to the development of symbolic coding, and ultimately, to cultural institutions. Sociality was entwined in virtually all these aspects of hominid life. Thus it seems impossible to separate the brain and its mental emissions from the various methods by which hominids responded to habitat heterogeneity. Some of these methods—more complex spatial maps of resources, for example—were tried nearly 2 million years ago, while others, like complex symbolic coding, were displayed rather late in the Pleistocene.

The early australopiths had a brain volume of about 400 cubic centimeters, very close to the size of an ape's brain. This was already large by the standards of other mammals, but it was to expand another 1,000 cubic centimeters in the lineage of *Homo.* Only the last 40 to 50 percent of the increase was strictly independent of body size. It is likely that the largest braincases of early *Homo,* and certainly the braincases of early *Homo erectus,* came attached to taller and more massively muscled bodies than their predecessors'. By the time of *Homo erectus,* hominid bodies had evolved to modern proportions. From then on, most of the increase in brain volume was unrelated to the body and given over to cognitive functions. Virtually all of this increase occurred within the past 1 million years.

Brain size was susceptible to variation, evidently a malleable trait in early hominids, but it was not until environmental fluctuation began its last, long crescendo that mental pressures took more or less exclusive control. It was at that point, I surmise, that our neural organs began to serve the needs of flexibility. The prehistoric record of hominid behavior gives no direct reflection of the mind's flexibility. It is mute on the subject of Paleolithic imagination. But we have seen hints in the buried residues of the past. The open programs special to the human brain, which invite novelty and complex cultural behaviors, were written in the time most swayed by environmental oscillation.

The evolved commitments of most species typically end in specializations. But in the hominids, the first small adjustments to environmental variation seem to have led to larger accommodations. The commitment to buffer and to adapt to environmental fluctuation presented a new path, which the genus *Homo* took. While some hominid populations latched on to specific types of habitats, the agenda of a generalist proved to be more successful in the long run than that of the specialist as environments oscillated with greater amplitude.

Some populations of hominids were lured farther and farther into the arena of a mutable natural world. A two-legged toolmaker, already the owner of a relatively large brain, may have had little choice but to accept the commandment to endure vicissitude. In the surviving lineage, the hominid brain became tightly bound to the sociality of the home base and the high-rise part of the Monolith. Together, these served the causes of resilience, novelty, and responsiveness.

The distinctive character of the human brain impels us not to contemplate our origin as a refining adaptation to a specific set of conditions, or as a progressive gain over a static natural world, but rather as a process of adaptation to nature's contingencies. The oversized appearance of the brain is a bold badge, reflecting the gauntlet of environmental change through which the ancestral path of *Homo sapiens* has passed.

10

Dispersal and Extinction. Two last points of business in the origin of our species are duly implicated in our present transactions with the natural world. The first is the movement of hominids over wide areas, the eventual spread from Africa across the tropical and low-temperate latitudes of the Old World into frigid zones, the New World, and the ocean islands. Dispersal into uninhabited places is responsible for the worldwide scope of human ecological influence today.

Humankind's geographic excursion appears to be in utter contrast to the rest of nature. Like intelligence, culture, and technology, the global spread of humanity seems to be part of our domination over environments, rather than part of our origin in the natural world.

Let us consider an alternative. The immensely effective dispersing and colonizing tendencies of human beings make sense as a product of evolution in Pleistocene environments. The ability to survive displacements from home turf, to track familiar resources over long distances, and to recolonize old areas were all favored as Pleistocene landscapes were revamped. The inclination to colonize new regions required only an amalgamation of the two basic responses to environmental fluctuation: to follow vital resources, which requires mobility; and to expand the life conditions in which survival is possible, which requires flexibility and ways of buffering change. Over the past 1 million years, hominids became masters of both.

Italian paleontologist Augusto Azzaroli has described the Pleistocene of Europe as a series of "dispersal events." Animal communities were revolutionized by intercontinental migrations and the replacement of one species by another. These events, he proposes, were triggered by climatic change and tectonic movements in the mountain chains of southern and central Asia. According to Azzaroli, the dispersal event by far the most prominent took place around 1 million years ago.

The blesbok of Lainyamok spanned the temperate zone of southern Africa and the equatorial rift valley in a grand sweep some 350,000 years ago. On several occasions, the warm-weather hippopotamus moved into northern Europe. Yet these occasions were interspersed with arctic conditions, indicated by species like the musk-ox and the collared lemming. The geographies of mammals, big and small, were susceptible to large-scale expansions and contractions. In this light, hominid colonization of new habitats was no more than an extension of events and struggles going on in the biotic world at large.

Taking the distances between archeological sites and stone outcrops as a guide, we can calculate that the early toolmakers of Olduvai, 1.8 million years ago, traversed regions of perhaps 300 square kilometers, with a home range usually in the neighborhood of no more than 30 square kilometers. At Olorgesailie, 1 million to 700,000 years ago, hominids traveled up to 46 kilometers to obtain raw materials, suggesting a territorial range of at least 6,600 square kilometers. And by the late Pleistocene, ranges of at least five times this area were not uncommon. The human presence spread as the vectors of landscape revision and amplitudes of oscillation grew stronger.

Sometime after 200,000 years ago, hominids occupied for the first time the northern plains of Europe, the rugged highlands of central Asia, the forests of central Africa, and even later, the steppes of the Ukraine. By 50,000 years ago, human populations were prepared to make revolutionary expansions to new continents—Australia, North and South America—and to ocean islands, vast sand deserts, and tropical rain forests. Ventures into very cold regions (e.g., Siberia) may have occurred before this time; but regular occupation of polar zones was delayed until perhaps 30,000 years ago.

Mainstream anthropology consistently ascribes this colonizing venture to new abilities that reflect the separation of human beings from their surroundings. The evolution of modern humanity, we are told, has largely entailed "being released from the shackles of environment by major developments in culture."

But the journeys from tropics to poles, from one continent to another and across the oceans, suggest a different view. While humans are independent of any single type of environmental regime, human mobility and culture were shackled to a natural world susceptible to periodic disruption. These facets of human existence were tuned to the signal of habitat extremes over geologic time. Human colonization reflected not a release from the environment, but a response to its long-term effects. *Homo sapiens* thus gained the fundamental properties of an organism conceived by a natural world rife with change.

A species meets its demise when the area it occupies dwindles to zero. To occupy a single sedimentary basin, or the confines of a single region, does not therefore give a species much cushion against the risks of extinction. When *Homo erectus* spread beyond the rift valley, the possibilities of complete failure were greatly diluted. By becoming a traveler, *H. erectus* happened to minimize the risk that all populations of his kind would vanish. This may have been a significant reason for the endurance of this species over a million years.

But the new residences *Homo erectus* found were as turbulent as those that African populations experienced. The whole of eastern Asia was affected by vertical displacements of the land and shifting habitats. According to Chinese researchers, the rise of the Tibetan plateau changed the basic pattern of the natural landscape established several million years before. Intense vertical movements over much of the eastern portion of Asia transformed a relatively homogenous, low-lying landscape into one of sharp relief during the Pleistocene. The uplift of this vast region magnified the oscillating strength of the monsoon that brought cycles of moisture and drought to eastern Asia.

Pleistocene habitats of Southeast Asia are less well documented. But there, too, the landscape changed dramatically as sea level rose and fell. The shifts were worldwide, nearing one hundred meters during some glacial-interglacial oscillations. A large proportion of peninsular Asia and the Malay Archipelago experienced dramatic inundations and exposures of terrain. During the last glaciation, 2 million square kilometers of the Sunda region, currently submerged, became exposed. Dry land connected the Malay Peninsula with all the islands to the south and east. Such vast episodes of land-sea fluctuation surely affected the size of early human populations in this region and, in the long run, the persistence of certain ways of life over others.

In every place where human populations lived, the fossil record shows that many animals became extinct. Twenty-seven species of large mammals are known from the rift valley of southern Kenya between 1 million and 600,000 years ago. Forty-eight percent of them are known to have become extinct by 350,000 years ago. This compares very well with the more abrupt turnover of species in Europe. Around 1 million years ago, more than 50 percent of the previously known mammal species disappeared without descendants.

In Java, where fossils of *Homo erectus* are well known, the biota underwent even more dramatic change. About thirty species of large mammals have been identified from the mid-Pleistocene fossil record between about 1 million and 200,000 years old. Twenty-four of these species—80 percent of the known fauna—did not endure beyond this upper time limit.

Nearly two hundred large mammal species are known from the early and mid-Pleistocene of China. About 70 percent disappeared by 150,000 years ago, and virtually all of them are now extinct. Certain groups of mammals left a single survivor. At least ten species of wild pigs lived in China during the Pleistocene, but only the wild hog *Sus scrofa* endured. In relatively recent times it inhabited an enormous zone from Southeast Asia to Siberia, across Europe and into the Sahara—certainly the most widespread and probably the most adaptable pig of the lot. Some groups of animals left behind not even a single species past the mid-Pleistocene. Twenty-six species of fossil proboscideans—elephants and related forms—are known from the Pleistocene of China, but not one of them persisted into recent times, an extinction rate of more than ten species per million years within just this one group of mammals. The living Asian elephant made a brief appearance in southern China late in the Pleistocene and is, in fact, the only proboscidean to survive in Asia.

The fossil record of China harmonizes remarkably well with findings in the rift valley of Africa, where biodiversity was streamlined over the past million years, often leaving a single lineage as the survivor in each group of large mammals.

It has always been tempting to ascribe the extinction of large mammals to human hunting. The division between natural and human causes of extinction pays tribute to the wall that an ecological account of human origin is slowly beginning to disintegrate. Belief in a stable, if not static, natural world means that the finger of guilt for high rates of extinction must be directed at humans. For some archeologists, the discovery of any environmental change or loss of species is considered an automatic marker of human presence. Supported by the main observations of our investigation, we may wonder whether the strict division into human versus natural causes makes any sense.

As far as we can tell, hominids had little influence on their environments prior to 100,000 years ago. Although tradition portrays *Homo erectus* as a big-game hunter and master toolmaker, there is little to suggest that he practiced predatory skills beyond those of his earlier ancestors. Nor is there evidence that his activities led to the extinction of any species.

Olorgesailie is one source of the idea that *Homo erectus* systematically killed for a living, a view largely based on a single excavation where thousands of bones and teeth of the giant gelada monkey were unearthed, together with hundreds of stone handaxes. Yet after diligent search, not a single convincing butchery mark or stone-tool percussion mark has been found on the monkey bones. Argon dates on volcanic ash and magnetic reversals in the sediments show that several tens of thousands of years were involved in the accumulation of bones and tools in the old stream channels of this site. Moreover, every other site at Olorgesailie in this time range also shows the natural predominance of giant geladas over other mammals. Lacking any evidence of systematic butchery of the bones, there is surprisingly little to justify the belief that toolmakers systematically hunted these baboons.

Archeological sites elsewhere suggest that *Homo erectus* collected a variety of bones and carcass segments from many different kinds of animals, which may mean that *erectus* practiced both hunting and scavenging, just as virtually all large carnivores do today. This current line of thinking is far from the old portrait of this hominid as a proficient, specialized predator on the verge of changing the balance of nature. The archeological record offers no sign of systematic, cooperative big-game hunting until perhaps 200,000 years after *erectus* became extinct.

Animal extinctions during the age of bipeds far preceded any hint of significant toolmaking or meat eating. The first hominids emerged from a biodiversity decline at the end of the Miocene. Later on, when the first human toolmakers appeared, mammal diversity in Africa again reached a high. But the sorting and culling of species took over once more as habitats oscillated widely during the mid-Pleistocene.

Repeated change in the landscape was one cause of the biodiversity crisis near the end of the Pleistocene. As newcomers to the Americas and Australia, humans also asserted a powerful influence on mammal populations. But it has always been difficult to separate the effects of climate and of human predation on the extinction episodes of that time. The splurge of human hunters over new landscapes occurred against a background of major climatic change, a time of ecological trauma in some places, such as Australia, and of ecological recovery from glacial conditions in others, such as North America.

The most challenging point about extinction is that its effect on the various lineages of hominids was just as harsh as it was on other mammals. The hominids, of which we are now the lone representative, were by no means exempt from the rule of extinction.

In a large sense, this is the cruelest reality of our origin. It introduces humanity, in its many faces and transfigurations over the course of time, to the chance of its own demise. Based on recent studies, even conservative estimates acknowledge that at least eight different species evolved in the human family tree over the past 4 million years. Most authorities recognize between ten and thirteen distinct species of bipedal hominids in this span. Perhaps only four of these were directly involved in the descent of living humans. Even in the link between ancestor and descendant, innumerable populations of the earlier species seem to have carried on for some time and ultimately left no descendants.

With each excursion into new terrain, populations of *Homo erectus* diversified and disseminated over a vast area a number of distinctive trials in human appearance. Differences arose between Asian *erectus* and African *erectus.* Some researchers have argued that these differences indicate divergent lines of descent, giving the new name *Homo ergaster* to the African fossils. Although I currently disagree with this practice, the regional diversity of *Homo erectus* is certainly greater and perhaps stranger than previously thought.

In eastern Asia, intriguing clues of variations within *Homo erectus* have been discovered. In 1989 and 1990, two skulls were found at Yunxian, Hubei Province, that show that certain populations of *erectus* varied measurably in the size and form of their teeth. One Yunxian skull possessed a last molar the size of a small peg, while another had cheek teeth as massive as those of the robust australopiths. For years, Chinese paleontologists have claimed, on the basis of isolated molars, that the large-toothed australopiths once existed in eastern Asia. Now it appears that those teeth were attached to *erectus* heads. The *erectus* population from Yunxian represented a distinctive gene pool, and showed variations not found in other fossil samples classed in this same species, such as those of Peking Man. Whether the Yunxian population was on the direct ancestral path of modern humans is unclear. What we know is that certain variants and combinations of features were ushered toward the future gene pools of human beings, while others were turned away.

Between 500,000 and 200,000 years ago, populations continued to diversify. According to one contingent of scientists, all hominids in this period belonged to a single evolving line connecting *Homo erectus* and modern *Homo sapiens.* Other researchers reconstruct as many as four separate species of human in this interval: *Homo erectus,* still alive in eastern Asia; *Homo heidelbergensis* in Europe; poorly known populations of archaic *Homo sapiens* in Africa; and toward the end of this time range, the famed Neanderthals, who had evolved in Europe and western Asia.

The Neanderthals inhabited the cold temperate zones of western Eurasia between about 200,000 and 30,000 years ago. Mainly because they buried their dead, they left behind a reasonable sampling of fossilized bones and partial skeletons. From these finds, anthropologists have discerned the unique Neanderthal anatomy, which has long prompted debate about these ancient people. Neanderthals had large brains, close to or even surpassing the modern human average, in part because they possessed such massive bodies. Still, their braincases retained an archaic appearance, long and low. The nasal and cheek area of their faces protruded, drawn out from beneath a prominent bony brow. Although variable, the front of their jaws tended to slope backward rather than jut forth in the familiar chin seen in people today. The entire tooth row was set forward, leaving a gap between the last molars and the part of the lower jaw that rises up to meet the cranium.

Below the jaw, their bodies were rugged. Their skeletal elements were fortified by thick coatings of dense bone, enabling them to endure great amounts of muscular stress. Raised, angular markings testify to powerful muscles and radically strong bodies. In contrast with the lean, equatorial body of African *Homo erectus* almost 1.4 million years earlier, the Neanderthal body was heavy and broad, estimated to be about five feet four inches tall and minimally 140 pounds in adults, or perhaps much closer to 200 pounds.

The Neanderthals inhabited the climates and terrains of Ice Age Europe and the Near East, environments to which they seemed specifically adapted. Their overall body proportions and relative lengths of arm and leg bones were indicative of populations with a long history in cold climates. Their enormous nasal regions projected forward, due largely to the backward sweep of the cheekbones. While related in part to tooth function, one effect of this process was to create a nose with an even greater internal volume than that of earlier, tropical hominids. This, too, would have proved effective in warming and humidifying air drawn into the lungs—a difficult but vital function in very cold climates. Their bodies, faces, and geographic distribution all suggest that Neanderthals were built to withstand the cold habitats of late Pleistocene temperate zones.

The skeletons of old and injured individuals show that the Neanderthals maintained and perhaps greatly cared about the lives of elderly or infirm members of their societies. They buried their dead. They incised bone and stone in a way that had symbolic meaning, though the frequency, standards, and variety of their symbolic activities did not approach that of later, modern peoples in the same regions. In the total scheme of things, they made simple tools of stone, essentially the same tools made by the earliest modern-looking people. It has even been suggested that late-surviving Neanderthals could repeat techniques of stone knapping invented by modern humans who migrated into Europe around 40,000 years ago.

The Neanderthal and modern peoples shared certain ways and probably also a large number of genetic variations due to common ancestry. At the same time, both populations maintained their separate identities, captured in their bones.

A pelvis discovered in 1983, part of a well-preserved Neanderthal skeleton from Kebara Cave, Israel, illustrates in a single bone the package of traits that lived and died out with the Neanderthals. The blades of the hipbone were splayed out to the sides noticeably more than yours or mine, meaning that the hip sockets faced less forward, which probably had a subtle effect on the way they walked. The pubic bone was a long, thin bar, unlike the brief, compact span of bone in your own pubic region. The rim surrounding the birth canal and urogenital organs had a distinctive heart shape, indented strongly from the back and elongated in front. Before this find, the main evidence of Neanderthal pelvises consisted of several pubic bones, all showing the elongated bar. Now we know that those bones were part of pelvises different in overall appearance from those of modern humans.

The Kebara find suggests the singular combination of traits that distinguished the Neanderthals. From the attachment of muscles on their shoulder blades to the peculiar proportions of their fingers; from the distinctive midface thrust to their personal bony massiveness, the hardened core of the Neanderthal body had a mannerism all its own that has not endured.

This unique skeletal amalgam persisted for a long time, even after modern-looking peoples evolved. The period of overlap in the Near East was at least 50,000 years, and in Europe perhaps 10,000 years. Certain Neanderthal skeletal traits appeared in some of the modern skeletons; but these are found in isolation, not joined in the federation of traits specific to Neanderthals. Nowhere in the thick alphabet soup of modern genetic diversity can the unique combination of Neanderthal traits be reproduced. In this sense the Neanderthals are an extinct form of human being, outsurvived by a different experiment in humanness, modern *Homo sapiens.*

The dismantling of the Neanderthal gene pool was by no means automatic. Over tens of thousands of years, Neanderthals and populations of more recent appearance apparently swayed back and forth across the landscapes of the Near East. According to the discoverers of the Kebara pelvis, Ofer Bar-Yosef and Yoel Rak, Neanderthals may have prevailed during periods of colder climate. But by around 40,000 years ago, something had allowed populations with the surviving combination of traits—rounder heads, chinned jaws, longer bodies, and slightly different limb proportions—to colonize new zones. The demise of the Neanderthals coincided with this final burst of human populations into new lands and environmental domains. The spread of new traits also enveloped western Europe, which had been Neanderthal territory for at least 150,000 years.

There is a popular notion that cold, glacial habitats fueled the final stages in the emergence of *Homo sapiens.* If this was so, how is it that the consummate Ice Age specialists, the Neanderthals, became extinct? The quandary here is similar to the problem of the robust australopiths, whose special dental adaptations to the vegetation of dry, cooler conditions failed to assure their survival in the savanna. It doesn't make sense that the Neanderthals became extinct under the very conditions that generated their unique adaptations in the first place.

The problem lies less with the Neanderthals than with certain assumptions about evolution. Natural selection acts generation by generation to create the match between the gene pools of organisms and specific habitats. In this way, the individuals comprising a species inherit certain genetic adaptations. Applying this idea to human evolution, our tradition has been to identify the special—usually the most challenging—environment in which the proclivities of our kind came to fruition. To many authorities, the shorthand answer resides in the cold winds of a glacial world.

But do we really mean that the dynamic tendencies of the present human species accrued in any one type of ancestral habitat? In this chapter, I have presented the reasons why a different view of the environmental milieu of human origin needs to be embraced. Suited to a particular geography and range of conditions, the traits special to Neanderthals are precisely those we might expect to have been replaced by behaviors and skills honed to accommodate the long-term extremes of environmental variation. The Neanderthals were a lineage of successful, long-lived people. But there is nothing to indicate that they begat populations of human beings who colonized the vast climatic and habitat diversity of the world after 50,000 years ago. Eventually, the peoples whom we in our bias call "modern," simply because they were the ones to endure, expressed an extraordinary degree of adaptability—sufficient even to replace the Neanderthals in their own special zone.

II

In the opening chapter of this book, I asked you to consider the lives and extinctions of human predecessors. It is very hard to evoke deep feelings about a strong-browed *erectus,* or personal attachments to the immensity of a million years. Yet the prevailing ecological dilemmas of our time must draw us personally to the fundamental question of our study: How could nature have produced a species like *Homo sapiens?*

If the natural world is as we've long envisioned it with a stable, definable original state, from what part of nature are human beings derived? And if our origin takes its rightful place in a natural genealogy, a kinship shared with all other living beings, how is it that we alone rub so destructively against the stolid grain of nature, the regularity and evenness of habitats to which organisms are adapted, the beautifully blended communities of plant and animal species that we have long believed to exist?

If nature in the absence of human beings is firmly established and in balance, why has the surviving hominid gone to extremes to create personal and cultural security? Modern human society is now urged on by government programs and amazing global connections that attempt to assure the status quo of industrial powers, worldwide economic distributions, nations, paths to social status, and methods of extracting resources from Earth and maintaining their flow—not just through our bodies, but through the very matrix of our institutions, which must be sustained, it would seem, at any cost.

There are several simple theories derived from facts about the natural world that seem to explain these human tendencies: the selfish gene; the need for status; the territorial imperative, among others. Each of these theories has sound reasoning. But each has been nourished on the assumption that the environment of human origin was a stationary target against which hominids over time improved their adaptive and competitive edge. Thus one progressive foothold after another was gained, leading to the self-centered exploitation of the world.

In the end, we must acknowledge that living humankind originated in a natural world quite different from the one we have commonly constructed and that serves these theories well. Human existence required surviving in an arena of nature where organisms are responsible for their own assurance that the world will continue tomorrow. It is a place of unexpected, indifferent variation, a world that, if a lineage is to survive, its members will ultimately be called upon to buffer, to create a network of options oriented toward developing some modicum of stability.

In a species that becomes especially good at this exercise, the modicum of stability begins to serve as a worldview. The baseline rises to a level that instills a sense of permanent security. This sense of the world is, however, false, and its ideal about how the world operates has the power of myth.

The visions that opened this chapter also possess that power, and we may now respond to them:

Alterations of great scale have been part and parcel of the most recent geologic epoch on Earth. Broad regions, often entire continents, were affected by environmental change. The rate and type of impact have varied around the globe, and the degree of alteration has increased sharply over the past 1 million years, yielding perhaps the largest amplitudes of habitat change in Earth's history. All this occurred before any hint of human dominion.

Alterations were not always gradual. Toward the end of the Pleistocene, nature's changes matched the rapidity of human disruption observed over much of the world today. Ice Age fluctuations between glacial and near-interglacial conditions sometimes occurred in a century. And in some places reversals between forested and open biomes were completed in less than that.

The baseline rate of extinction has always been portrayed as very low, consistent with natural stability. In contrast to the Miocene acme in mammal species diversity, however, the period of human origin was hit very hard by extinction, even before the beginning of the current crash in biodiversity 10,000 years ago.

The present ecological dominance of human beings cannot be questioned. The elements on which this dominion is founded were, however, pieced together over a very long period of time. The bipedal forebears of modern humans all possessed some combination of these elements. But this did not ensure their survival in Pleistocene environments, or in the current biota. Extinction was as forceful in molding the history of hominids, including the lone survival of *Homo sapiens,* as it was on many other large mammals.

Technology, culture, the innovative use of resources, and a worldwide presence now underwrite the ecological impact of our species. But the development of these special characteristics need no longer be deemed aberrant and in conflict with the natural world in which they arose.

Their primary place in the heart of humanity's repertoire reflects the mutable path of past environments. They are a mirror of nature's quicksilver. The natural world has emitted a creative dimension in time that belies our static measures of its present habitats. These reflections on the long environmental history of our species lead us to a new view of nature and of our relationship with it.

CHAPTER VI

A NEW VIEW OF NATURE

I

EVOLUTION has two fundamental bases: First, all living things are connected by descent; second, environmental context shapes the history of life. Earth is surrounded by organisms that replicate, breathe, and maintain themselves. The structure of the biosphere and the character of the species in it are a legacy of the planet's curious environmental path from past to present. The way organisms live, what we think of as their *nature,* reflects the conditions under which they evolved.

What is "the natural world"? Has our investigation so far changed our definition of it in any way? We popularly envision nature as an untouched meadow, the waving of tall grass, a silent woods, a mountaintop, or an unlittered beach. It is what we see and breathe when we are far from urban cement. Nature is composed of particular environments that can be inhabited or visited—landscapes, colors on a map, observable boundaries, heat and humidity that can be measured, solid memories both pleasing and troubling.

But there is another side, which would be impossible to know if we couldn't explore beyond our personal limits. Nature is a saga with a past, present, and future. The world is not composed simply of natural states, but of their motion through time. The sensed, measured, depicted aspects of any habitat are actually moments of illusion, photographs taken while nature is in the midst of movement.

Environments consist of states; environments are mutable. Both statements are true, but it is the former that governs our view of nature. The idea of nature's inherent stability has the power of hypnosis, and our trance persists despite shaking discoveries about the mutability of genes, species, and climates.

Stability means one of two possibilities: an environmental standard signifying nature's original condition, or a propensity to return to a particular state even if change has occurred. In either sense, stability consecrates nature as an original, permanent essence. These two narrowing concepts of nature's status quo pervade public beliefs, philosophies, and understandings; they are the great lenses through which we envision the natural world.

While ecosystems are resilient, Earth's odyssey undercuts the notion that there is one state to which nature continually reverts after disruption, or even that there are two states between which environments continually bounce. Using the Pleistocene as a guide, we see that nonlinear change has etched intricate paths of environmental alteration, not an orderly, limited pendulum between wet season and dry, cold and warm, but a bizarre course where every environmental factor magnifies, diminishes, or puts a novel slant on the way the world used to be. The mutability of environments and organisms cannot be disputed. A new concept of habitat change must color our thinking about our surroundings.

In the 1870s, geologists began to accept evidence of the buildup and withdrawal of huge glaciers over the northern continents. In 1909, German geologists Albrecht Penck and Eduard Brückner proposed that the Ice Age of the Alps consisted of four periods of ice-sheet growth, interrupted by periods of warmth.

In 1955, Cesare Emiliani of the University of Chicago published the first intricate zigzag curves of climatic fluctuation, based on studies of oxygen isotopes in the shells of forams. By 1969, Nicholas Shackleton of Cambridge University had shown that Emiliani's curves reflected tremendous oscillations in glacial ice and global climate. As these studies proceeded, geologists working in Africa proved that the continent's environmental history was far more complex than the old fourfold division of Pleistocene climates.

During the 1970s, teams of researchers dissected the patterns of Pleistocene fluctuation and exposed the many turns in the isotope curve. The long-standing division of alpine climates was replaced by a more complicated scheme. In the 1980s, sequences of fossil pollen and oceanic cores penetrated deeper than ever before, illustrating the long history of oscillation and its fantastic crescendo in the later phases of geologic time.

We have known for some time about the Pleistocene's climatic fluctuations and might question whether a new view of nature is really justified by what we have discovered. But the ideas of stability and progressive change are deeply rooted in our popular and philosophical perceptions, and sharply reflected in the biological sciences. The framework of biology mirrors our present understanding of the natural world and our relationship to it. Our view of how living things evolve and adapt in the natural realm takes us to the very heart of science's concept of nature.

Natural selection is widely considered the fundamental process of evolutionary change. Darwin's concept makes sense out of both the beautiful and the jerry-built structures of organisms and explains much of what is useful and odd, beneficial and antagonistic, in their behavior. It has spawned new observations about the development and genetics of living creatures, and it has been tested by models of population biology, game theory, and biomechanical design. It has also been challenged by paleontologists studying the tempo of change in fossil species. Through all this interrogation, the concept of natural selection has remained at the center of every field involved in the study of living things.

Genetic variation between individuals is vital to natural selection. DNA is a code not unlike a computer program, but with chemical instructions that affect maturation, appearance, physiology, and behavior. Variations in DNA influence survival, mating, fertility, and success in rearing young. The word *selection* simply means that in the face of the environment, certain variations lead to higher rates of reproduction and survival of offspring. In Darwinian selection, the environment is a sieve, influencing the fate of individual variations and therefore the makeup of the gene pool in future generations.

The case of selection most familiar to biology students is protective coloration in the British peppered moth (*Biston betularia*). This moth has two appearances, peppered light gray and very dark gray. Both forms are preyed upon by birds. Being eaten depends upon the moth's visibility against tree trunks where it alights. The peppered appearance blends extremely well with the mottling of lichens that cover the tree trunks in the unpolluted countryside. The darker variety avoids detection where tree trunks have been blackened by factory soot; against this background, the peppered form becomes both conspicuous and food.

The simplest of genetic variations controls the two appearances, each of which has an advantage in one of the two environments.

Darwin saw reproductive success as the outcome of what he called "the struggle for existence." But the struggle need not involve head-to-head battle, as the moth example illustrates. In most cases, it is simply a matter of limited resources, which restrict the number of offspring that survive. Any species has certain general qualities, propagated through time, that are found in virtually all of its members. This does not mean that survival and mating are equal in all individuals. The general attributes are transmitted from one time to another by streams of individuals who happen to reproduce successfully. They carry the general attributes of their kind and their own particular variations of the DNA blueprint, which result from mutations and new mixtures from the parents' pool of DNA.

Natural selection is the fate of these variations, the influence of the environmental conditions at hand—food, water, shelter, mates, predators, and so on—on the pool of genes relayed to the next generation. This influence is manifest at many different levels, from the social, where individuals affect one another's reproductive success, to the molecular, where genes affect the function and inheritance of other genes. Darwinian selection is often discussed as if it were simply the difference between individuals in reproductive success. But mere comparison of the number of offspring ignores the milieu, which Darwin made the critical determinant of selection.

Darwin's analogy to animal and plant domestication clarifies this idea. By selecting certain variants over others, breeders are able to change irrevocably the appearances and reproductive abilities of domesticated animals and plants. Variations also arise in nature as a result of reproduction. So, Darwin asked, did nature also possess a way of honing these variations and permanently altering the characteristics of organisms?

Darwin and his contemporary, Alfred Wallace, scouring independently in Malaysia, established the mechanism of natural selection: As the environment changes, the conditions in which organisms live are biased against certain variants and favorable to others. The direction of environmental change creates a consistent effect over a long time, which allows the development of new behaviors and other means of adaptation. As a result, offspring come to differ from their ancestors in the long run.

Three pillars of the Darwinian concept of selection show how biology perceives the natural world. The first is the generation-to-generation time scale of selection. The breeders analogy led Darwin to contemplate a slow, inexorable tempo of change. Natural selection acts by culling and sorting variations in each generation. If a new variation improves the survival and reproductive potential of its recipients, it increases in the population over time. Generation by generation, it will be put to the test. Its persistence in each generation is the currency of its success.

The second pillar involves the consistency of selection. Darwinian selection is a process by which new variations improve growth, survival, or reproduction in the presence of a particular environmental signal. A given intensity of sunlight is one such signal, governing which differences in the sensitive retina will prove most helpful to an animal. Rainfall and water availability are another, determining the variations of kidney function that best promote survival. They also affect the time an animal spends looking for water, and thus the time devoted to the search for food or mates. Organisms are also signals to one another. As the orchid affects the way in which bees gain access to nectar, so the bee influences the best orchid designs to assure pollination.

According to this fundamental vision of biology, evolution is harnessed to the specific status of the habitat. Biologists talk about the relationship between an organism and "its environment" as though the latter were a single thing—a state to which the organism is matched. Thus the special adaptations of any organism reflect a specific environmental state that determines which variants will succeed over others.

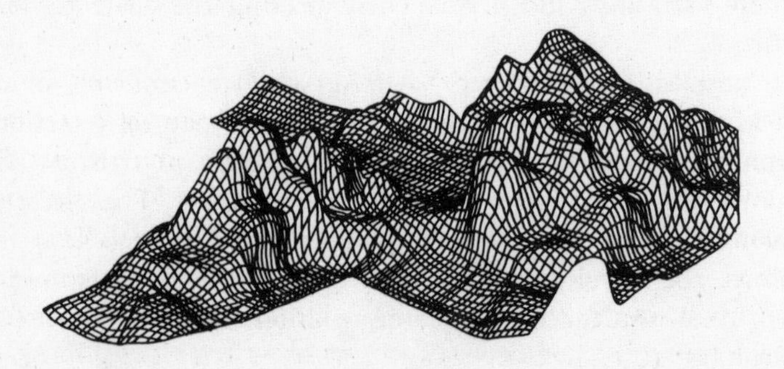

One of the twentieth century's most powerful images of evolution is the "adaptive landscape," developed in the 1930s by geneticist Sewell Wright. All possible combinations of genes in a population or species are laid out horizontally across the landscape. The topographic relief, defined by peaks and valleys, depicts the degree of fitness—that is, the relative advantage of each gene combination in meeting the demands of the environment, an advantage that determines which genes are passed on to future generations. A peak indicates success; a valley, a dead end. The tendency of certain gene combinations to move uphill over time represents the process of adaptive change.

This process is possible only if the topography remains relatively constant. A population that varies its gene pool slightly in every generation must experience consistent conditions for selection to push it uphill to a viable place on the adaptive landscape. Evolution by Darwinian selection is the ongoing match between a population and its consistent habitat.

The third pillar is the fact that change occurs in response to a new environmental direction. Adaptation is thought to involve a long-term process of response to a particular direction of environmental change. In Sewell Wright's landscape, the movement of adaptive peaks in one specific direction makes Darwinian selection a process of ongoing creativity and change, not a process of perfection.

Biologists commonly say that organisms are subject to *selection pressure,* which means that environmental change "forces" the gene pool in a particular direction. There are mathematical ways of expressing this force, one called "relative fitness," another, "the coefficient of selection." These have meaning only to the extent that the conditions of selection act in a consistent manner over time, retaining certain gene combinations at the expense of others.

Another scientific term that highlights directional change is *environmental forcing,* the idea that species' origins and extinctions are caused by major trends in the environment. As the environment is transformed from one state to another, populations may become extinct, divide into two, or adapt to the new situation, underlining the idea that environmental change in a consistent direction is crucial to evolution.

Basic biology texts often depict these three pillars of evolutionary thinking in bald terms. Biologist Niles Eldredge writes: "Long-term, directional natural selection . . . is the basic explanatory device of evolutionary theory." Without the slow, inexorable sculpting of Darwinian selection, it would be hard to imagine how evolution shapes the complexities of organisms. Each period of selection has to produce results in accord with all other periods. In current thinking, complexity arises from the congruous effects of natural selection over a long time.

Prevailing ideas about the evolutionary process give virtually no role to the most fundamental theme of environmental change over the past 6 million years: greater variability, frequent shifts in environmental direction, unexpected moments of change. Many living species have confronted a wide range of survival conditions. But biological theory has focused on stability, progressive environmental change, and what happens to individual organisms a generation at a time. It offers no clear insight into the ways in which the biotic world resolves the potential conflicts of natural selection experienced by *different* individuals of the same lineage over time.

The fluctuation between extreme conditions of survival has been treated as scarcely interesting, much less a significant evolutionary factor. *Extreme* must, of course, be defined in specific terms for each population of organisms. Extreme conditions of survival and fitness are those that test the limits of a population's persistence in any one place over a period of time.

In some instances, biologists acknowledge variation but believe that an organism adapts to the *average* condition of its environment. The *range* of habitat alteration is swept aside, and the functioning of an organism is not thought to mirror the intense signals of fluctuation endured in its past.

Many studies of animal behavior stress the uncertainty in the lives of organisms—the risk of being eaten, the competition for food, the conflict between time and energy. But animals face these matters of survival daily or seasonally, within a specific environmental sphere. Much of the satisfaction in the study of behavior comes from seeing how animals have adjusted to the habitats where we now find them, suggesting the importance of stable environmental states on our view of nature.

This idea of adaptation is at odds with the situation where environs are remodeled again and again over long periods. Since species etch their mark into the planet's archive over such spans, we can guess that extreme reversals of environment were also important in the evolutionary process.

Richard Levins's *Evolution in Changing Environments,* published in 1968, observes that short-term success has to be balanced against "the probability of long-term survival on a geological time scale." There are, he points out, "characteristics of organisms and populations which are not explicable as adaptations to particular environments. . . . These and other traits may be regarded as adaptations to the pattern of the environment in space and time, to temporal variability, to environmental uncertainty. . . ."

By focusing on variability from one generation to the next, Levins ignores truly long time scales. Although his ideas apply to complex species, he concentrates on the physiology of simple ones. Nevertheless, he and a handful of other biologists have seen beyond stability and directionality, the two forces on which science's concept of adaptation is based. Let us consider how organisms may adapt to long-term disturbance, and explore the implications for human origin and the present state of our species.

2

The conditions of natural selection shift periodically over long intervals. New characteristics that enable organisms to endure dramatic fluctuations in their surroundings are favored, and are shaped by disparities in Darwinian selection. Populations of organisms evolve the means to buffer extreme and repeated disturbances in the factors that govern their existence. I call this interaction between populations and oscillating environments *variability selection.*

Darwinian selection matches an organism to the norms of its environment, where innumerable aspects of its physical, biotic, genetic, physiological, mental, and social environments are consistently manifested. The vertebrate heart contracts and dilates in a certain manner, regardless of habitat. At any time, choosing a mate is affected by behaviors in the mate that best ensure the reproduction of the germ line. These facts of life exist regardless of variations in ecological settings and the resources of survival.

The various parts of an organism—genes, cells, anatomical structures, physiologies, and behaviors—operate together in ways that aid the organism to solve recurrent problems of survival, and to pass its inheritance on through offspring. Organisms also sometimes develop critical dependencies on other species. These are called co-evolutionary relationships and are assumed to take many millions of years to develop. When the African honey badger *(Mellivora capensis)* hears a certain song of the honey-guide bird *(Indicator indicator),* it means the presence of a beehive. The stout, vicious badger follows the song, tears the hive apart with strong claws, and, as he feeds, the bird also gets honey and insects as payment for services rendered. In all such matters of mutual benefit, the results require stability in certain life conditions—in this case, overlap in the ranges of bees, bird, and badger.

There are cases, however, in which extreme ranges of environmental variation test the strategies of an organism and the persistence of its lineage. In the paradigm of Darwinian selection, the functional designs of a species reflect the stable and regular aspects of its environment. The point of variability selection is that certain designs, behaviors, and ecological characteristics are fostered by discrepancies in the environment, caused by habitat instability over long time frames.

The basic assumptions of variability selection are:

First, each organism's genetic code is drawn from the gene pool of its population, which is shaped by random events involved in reproduction, and by a cumulative response to external conditions. All methods of survival that extend a population into future generations are the result of a lengthy history of response to these conditions.

Second, in order to persist, all organisms must be malleable to variation in their habitats. Seasonal changes or any other regular variation in environments experienced during a lifetime are part and parcel of generation-by-generation Darwinian selection.

Third, individual species may have long durations—in mammals, from hundreds of thousands to several million years. Over this span, a lineage may face wide differences in survival conditions, far exceeding seasonal variations.

Fourth, certain novelties are favored because they enable the population to endure the widest environmental disparities. These include any attributes that enhance the tracking of resources as they move, shrink, or expand; the ability to buffer totally new habitat conditions; and the division of a species into diverse, far-flung populations, thus lessening the risk of extinction.

Annual seasons, rainy-dry or warm-cold, are the biggest environmental changes most organisms ever experience directly. According to climatologists Thomas Crowley and Gerald North, "the change in temperature over North America from winter to summer is far greater than glacial-interglacial changes in mean annual temperature of the Pleistocene." This seems to imply that seasonal changes encompass most of the environmental diversity to which plants and animals are adapted. If this is true, it also means that the largest variations are met within each generation—the normal time frame of Darwinian selection.

Time has a way of playing tricks, however, as short periods pass into much longer ones. Yearly variations considerably exceed that of a season, and the climatic encounters of a lifetime are much greater still. If the history of Earth's environments tells us anything, it is that the *kind* of habitat disturbance changes dramatically with even longer passages of time.

Two separate rainy seasons in Africa produce the richly populated, short-grass savanna, while a single long rainy season produces taller, less digestible vegetation, incapable of supporting large herds. The annual cycle creates habitats of a predictable appearance: From one season to the next, trees remain rooted in place; lakes and rivers stay where they are; the boundaries between forests, deserts, and meadows are secure. The occupants respond to this relatively fixed world—what we usually mean by a "habitat." Their lives and environmental strategies reflect the routes to biological success in this particular setting. Even the narrowest specializations may evolve in response to the seasonal pendulum.

Let us run this image through a considerable span of time, where expanses of dry savanna give way to coalescing ribbons of forest, and then back again to open savanna, which occurred in the Pliocene of East Africa; where rivers and lakes appeared and disappeared, as archived in the geologic record of Olorgesailie; where dry, windswept plains alternated with moist woodlands; where habitat perturbation reared up again and again on a local-to-worldwide scale.

Expressed over thousands to hundreds of thousands of years, these movements and revampings of habitat are of a totally different quality from seasonal changes. The process of nature critical to my argument connects this extreme range of instability with the genetic continuity between generations.

As Pleistocene environs changed, certain organisms found it possible to follow the features of the landscape important to their survival. In many cases, they were even able to stabilize the factors affecting their existence—the types of berries they ate, the arboreal shelters they sought, the humidity and temperature they found comfortable.

For other organisms, the movement of biomes created new habitats and novel combinations of variables. As biologist G. C. Williams notes, in each environmental swing, Pleistocene populations faced distinctively new regimes of light, heat, and moisture, and new physical geographies at different latitudes. Since plant species migrate at different rates, vegetation changed from one episode to the next. In the same way, communities of mobile animals broke apart and reassembled with different members.

What we think of as "the same" forest or grassland comprised, from the organism's point of view, strange new variations. From time to time, populations had to deal with different arrays of competitors, predators, and parasites. The union of factors that favored survival during one period may never have recurred later. The prime resources of one epoch may have been lost or greatly reduced in another. Physical barriers limited the mobility of populations. Foreign conditions swept over the landscape, and the possibility of extinction loomed over any population unable to spread or migrate.

Even over a period of many such changes, organisms vie with one another, generation by generation, for the resources within relatively stable habitats. Selection in each generation may act swiftly to remarry the properties of the organism to each new setting. The genetic code for these properties is reoriented with each fluctuation, the gene pool of the population tossing and turning as it tracks environmental change. Genetic variation is one way of broadening any animal's response to habitat change over time. The replacement of one gene by another is helped by keeping a large storehouse of genetic variation within the population, a phenomenon widely found in nature.

Gene replacement can alter relatively simple traits in a vacillating setting. Quick or more extreme fluctuations may require a higher level of response. Under some circumstances, a premium is placed on developing a versatile path of maturation. As the individual grows, its appearance, physiology, or behavior is adjusted to outside circumstances. The population thrives without having the genetic code change with each shift in habitat. Repeated fluctuations over time may promote this process of acclimation, and larger oscillations may help to widen the range of possible responses.

Another kind of evolutionary response is possible where new methods of survival evolve from behaviors already offering some degree of flexibility. Novel and complex behaviors are built up over time in response to a wider and wider range of conditions, including unfamiliar challenges that require new strategies. This process may expand certain functions of the brain or enhance any aspect of social behavior and grouping that offers options to the organism.

Like all forms of natural selection, variability selection is the cumulative effect of an environmental signal. It occurs when the conditions of survival vary over time, and its effect grows with wider oscillations. In Darwinian selection, long-term change comes from adding up consistent advantages from one generation to the next. The results of variability selection, by contrast, cannot be deduced from the advantage fostered by any one specific environment. Instead, vacillations in the short-term advantage, governed by environmental disruptions over a period far exceeding a generation, are the deciding factors of biological success. The physical setting, ecology, and social relationships of one generation differ from experiences a thousand generations before and a thousand generations later; this sporadic reconfiguring of the short-term optimum is the basis of variability selection.

As the terrain is remodeled over and over, variability selection crystallizes any novel characteristic of an organism that helps it survive the composite of all environments. It is responsible for anatomical structures, behaviors, and strategies that benefit an organism in the contrast between densely forested and open habitats, long-term drought and water saturation, abundance and impoverishment.

Where the resources of survival are repeatedly changed, a population of generalists is favored. The difference between generalist and specialist is neither simple nor direct; it depends upon the factors governing survival and mating success, which vary from one kind of organism to another. By broadening the response to large, long-term fluctuation, the generalist gains the power to overcome disparities in the specific factors affecting its lineage.

The continuity of any lineage proceeds a generation at a time. So all genetic change must somehow be tied to the tempo of genetic inheritance within populations, the tempo of a generation. The largest fluctuations stimulate evolutionary change because Earth's environments are made up of innumerable cycles that vary widely in scale. As scientists are now aware, environmental change is a continuum that links the weather of a year to the more disruptive cycles over many thousands.

Exceptional warming of the Pacific Ocean happens every several years. This phenomenon, called El Niño, results from a complex exchange of heat between atmosphere and ocean. It induces torrential rains in one place and searing drought in another. It affects South America, Australia, Indonesia, and Africa, and it alters jet-stream patterns over North America. The effect on weather usually lasts for a year or two.

Sunspot activity represents a slightly longer period of oscillation. It is caused by variations in the solar magnetic field every twenty-two years, which upset weather patterns and possibly cause drought in the Great Plains of North America. Its exact effect is not yet well understood.

A recent study in the Kenya rift valley, just south of Olorgesailie, reveals even longer spans of environmental fluctuation. A 40,000-year record lifted from the bottom of Lake Magadi shows repeated shifts in the lake's chemistry, signifying climatic cycles up to thirty years long. They are believed to result from interactions between lower-level oscillations in seasonal climate, El Niño warming, and sunspot activity.

Using worldwide temperature data, a recent computer model of atmospheric conditions suggests that there is a 70-year cycle of global cooling and warming; while the tree rings of foxtail pines in the Sierra Nevada indicate a 125-year period of temperature rise and fall.

In deep-sea cores off the northwest coast of Africa, high-frequency fluctuations of about 600 years, with secondary cycles of about 1,150 and 1,700 years, are found in temperature curves for the late Pleistocene. During the last glaciation, flotillas of icebergs were discharged into the North Atlantic every 6,000 to 7,000 years. In the pollen recovered from the Grande Pile peat bog in France, significant changes in the pollen curve were found to occur every 8,800 and 9,200 years. A longer period of environmental variation, about 10,500 years, has been measured in the carbonate content of ocean sediments, a fluctuation apparently caused by an extra wobble in Earth's spin axis, which affects climate change mainly but not exclusively in the tropics.

These cycles ultimately link up with the largest fluctuations known so far—orbital cycles 23,000, 41,000, 100,000 years long—causing vacillations in the quantity of solar radiation, glacial ice, and tropical moisture.

Volcanic eruptions and uplift of mountains and plateaus exert a strong, erratic effect on regional and global environments. Climatic cycles and these more sudden seizures combine to create an extraordinary pattern of environmental change. The key point is that the various rhythms form a nearly continuous spectrum of change.

Variability selection requires a population to remain viable while facing a sequence of disparate environments. Genes that favor an ability to cope with seasonal cycles may confer a certain adaptability to slightly more intense fluctuations. Adaptation to decade- or century-long cycles may permit a degree of success in even more dramatic periods of change. Eventually, behaviors, mental functions, and social strategies that persist through wider oscillations lay the groundwork for meeting totally novel conditions experienced over even longer spans of time. In this way, adaptation to disturbance may be favored over adaptation to particular habitats.

Variability selection requires a particular kind of genetic variation. If the only variants available are ones that match the organism to a specific environmental state, no new means of flexible, innovative behavior can evolve. But other genetic variations favor open programs of behavior that vary and extend the adaptive possibilities of the individual. These are conserved in the gene pool over time because of inconsistency in the short-term effects of natural selection. Organisms eventually build up an inheritance that enables them to buffer larger and larger disturbances in the factors governing survival and successful reproduction. The chance that a population will survive an environmental change is raised by the buffering qualities its parent populations accumulated. Eventually, this process obviates the need for further genetic change in a totally new environment.

The spectrum of environmental variability links the happenings of a lifetime to the remodelings of a geologic era. The overall effect differs from the very rare, catastrophic impact of a large asteroid—what paleontologist David Raup calls "wanton change." Such disruptions, like the one that finished the dinosaurs, are indifferent to subtle distinctions between populations or species. Such devastating breaks limit the opportunities for the more patient evolution of the generalist. Instead, by the process of variability selection, lineages evolve on the numerous bridges of nature's alteration, merging the smallest with the largest.

Population diversity is a final factor in variability selection. Species are often represented by many populations, and any lineage prone to engender a diversity of experiments—variations on what it means to be a hominid, an elephant, a zebra, or a mouse—will have the best chance of enduring. Since populations vary, the most resilient are favored when adaptation to one particular habitat fails to assure survival.

Humankind, like other species, is the product of many time frames. Major shifts in environment occurred in chaotic episodes about 20,000 to 100,000 years long. The origin of our species was a thin thread slipped through the eyes of these longest needles of environmental change. At several places along the way, the branched path of our lineage outlasted alternative paths.

Our slender lifeline was eventually constructed out of a succession of *Homo erectus* gene pools and those of later archaic hominids. It faced repeated expansions and decimations until an accumulated pruning had culled the defining qualities of the bipedal, large-brained experiments now gone extinct, and had amassed the variations containing the possibilities of the only living hominid. This process was one that registered the conflicts, reversals, and shifting trends in the boundary conditions of human life.

3

In certain quarters of biology, the evidence in the previous chapters and my dissent about the ways of nature, evolution, and human nature will be quickly judged. There are two plausible objections to my theory. The first is what I call "the law of the minimum"; the second, and more serious, asserts that complex adaptations arise from interactions within the social group or with other species, independent of climatic variation.

Objection 1: It seems reasonable that natural selection operates by "the law of the minimum." Only times of drought, cold, and lowered food supply wield the sharpened scythe of natural selection. It is primarily during these periods that organisms are shaped to function in a particular way. Accordingly, human adaptation was honed in the drying trend of the African veldt and the Ice Age of the north, the harshest conditions in our ancestry. It is these conditions that sped human evolution and molded our character.

Although this belief is mainstream to many evolutionary biologists, it is a myopic view. It sees only the difficult times, periods of population decline, as relevant to the organism. But times of plenty are just as pertinent to the success of a lineage. During periods of population growth, diverse lines of kinship assert their differences in fertility. Novel experiments arise, probing the resource opportunities furnished during boons. Populations divide and disperse. The number of disparate gene pools multiplies. From these radiations, new behaviors may be tried that will later, under different conditions, affect survival. Such boons are of different but equal importance to periods of dearth in the persistence of a lineage through time.

The concept of buffering means that the population does not endure the largest swings from plenty to dearth to plenty simply by dedicating its genetic system to behaviors and relationships favored in one era versus the other. Instead, the genetic system is altered in ways that happen to improve an organism's versatility and response to novel conditions. The organism's possibilities are broadened, and the possibilities of the group and the species are diversified.

Ecologists have battled for years about whether competitive or noncompetitive interactions are more important in the operation of ecosystems. One side holds that animals are constantly vying for resources, and competition holds the key to both the cycling of resources and the evolution of populations in these habitats. The other side argues that animals actually avoid competition by taking advantage of different resources at different times.

These two viewpoints embody the discrepancies of natural selection—the importance of dearth versus plenty. Both are important at different points—times of plenty, when population growth, resource opportunism, and fertility come into play; and times of dearth, when bottlenecks, competition, and survival are key. Both assert their influence over a long span.

The idea of variability selection implies that certain populations of organisms succeed because of the conflict posed by this vacillation. Their genetic adaptations are not confined to one particular habitat; they are not geared specifically to dearth or to plenty, or to any one arena of social and ecological competition. They are designed to enable individuals to survive disparate habitats and resource distributions.

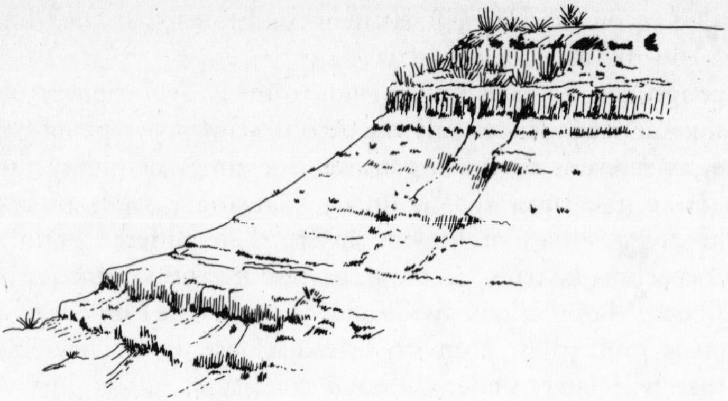

Objection 2: According to many biologists, neither climate nor any other factor in the physical environment is the chief instigator of adaptive change. Relations between organisms, especially within social groups, are what fuel the process of adaptation. As an individual gains evolutionary advantage, it influences nearby organisms. Over time, further changes heighten the relationship between coexisting organisms. As the prey gains a step in speed, a step gained by its predator will be favored. Further advantage to the prey is achieved when it becomes even quicker, and so on. Biologists call the overall process, which is a facet of Darwinian selection, an "evolutionary arms race." The result is cumulative, directional, and stimulates the evolution of complex behaviors.

In his book *The Blind Watchmaker,* biologist Richard Dawkins observes that evolution tracks changes in the weather, but that real evolutionary progress requires consistent pressure, not fluctuation, in natural selection. Since organisms are always an integral part of one another's environment, they provide the ongoing challenges and competitions that underlie the evolution of complex adaptations. The consistent presence of other organisms in the environment is, according to Dawkins, the reason genetic change and advanced adaptations are built up over time.

Arms races escalate the process between predator and prey, which eventually leads to extraordinary anatomies geared for speed, cunning methods of hiding or escape, and social cooperation in the hunt. Adaptation consistently advances because arms races tend to escalate in a particular direction. Dawkins emphasizes that such escalations have limits, and advancement may not be continuous. Still, he writes, "the arms-race idea remains by far the most satisfactory explanation for the existence of the advanced and complex machinery that animals and plants possess."

This presents a very serious objection to the concept of variability selection, and to the whole idea that the unique character of the human species owes its origin to large-scale contrasts in ancestral habitats. There is no doubt in my mind that evolutionary arms races occur and contribute vitally to evolution. At the same time, the environmental evidence of the recent geologic record documents the powerful theme of habitat disturbance.

The great merger of cycles and episodes affecting environmental change can no longer be considered mere weather, irrelevant to the interactions among organisms. Escalations of competition cannot be isolated from the vacillation in vital resources around which competition ultimately revolves. Resources are the foundations of habitats on which animal societies function, live, and die, so we can no longer ignore their movements and inconsistencies over time. Ultimately, organisms must be as adept in meeting the mutable and contradictory elements of their surroundings as they are with the consistent and progressive.

All field studies of animal behavior are conducted in specific habitats. Even in the longest periods of observation, the present environment is taken as the stable backdrop. Field observers want to find out exactly how the organism is adapted to its surroundings, how the baboon's social structure relates to its habitat, how best to define the ecology of finches. We want to figure out *the way* the lioness finds her mate and gains advantage over her competitors. We treat adaptation as a finely polished crystal implanted in each species, to be discovered at one field site or another.

To many biologists and popular writers, human nature reflects a very specific environment in the past, which some scientists have called "the environment of evolutionary adaptedness." Author Matt Ridley tries to define this in his book *The Red Queen,* the title referring to the ongoing arms race among organisms to maintain their adaptive place:

> Then for more than a million years [after the origin of *Homo erectus*] people lived in a way that couldn't have changed much. They inhabited grasslands and woodland savannas. . . . They probably lived in small bands; they were perhaps nomadic; they ate both meat and vegetable matter; they presumably shared the features

that are universal among modern humans of all cultures: a pair bond as an institution in which to rear children, romantic love, jealousy and sexually induced male-male violence, a male preference for young females, warfare between bands, and so on. There was almost certainly a sexual division of labor between hunting men and gathering women, something unique to people and a few birds of prey.

In this description, the environment is little more than a projection of modern human lifeways, inspired by observations of hunter-gatherers, and entirely independent of any actual investigation of past environments. It completely ignores the fundamental fact derived from studying the past: The social behaviors and biological nature of ancient hominids were caught in the flow of changing habitats. "The environment" was one of strong tides and large uncertainties. The character of the present human species became apparent only as environmental alteration became more and more powerful.

Direct scrutiny of the environments of human origin suggests that the rules of the evolutionary game were inconsistent over time. Wide-scale habitat change periodically altered the competitive milieu and changed the conduct of arms races. Changing the rules would have played a powerful role in the evolution of social exchange and ecological relationships. The unique and universal social attributes of humans, such as home-base behaviors and complex cultural systems, are sensitive to external variation. So it makes sense that the distinctive sociality of modern humans has been shaped more by discrepancy than constancy in the local surroundings.

The idea of a long, conformable arms race has also been applied to the evolution of human mentality. A significant development in the field of psychology views the human mind as a vast web of intricate mechanisms evolved as ways of solving specific problems our ancestors faced. These mechanisms are expressed in mental procedures, which are used time and again in human reasoning. The evolutionary psychologists, as they call themselves, tend to portray our ancestral environment as uniform over time, a view that results from two assumptions: that change occurs only by the consistent action of natural selection; and, that the evolutionary design of the human brain was largely a response to the social environment, causing an arms race of social advantage and problem solving.

In light of our findings, I would emend these assumptions. Change occurs in relation to the environmental *challenges* faced by organisms. In the long run, environmental instability posed the sharpest challenge to hominid populations, unexpectedly altering the problems of survival, competition, and factors affecting social strategies.

As the evolutionary psychologists have shown experimentally, many of our mental operations are shared with other social species and hold an important place in human affairs. But what about the reasoning processes that developed during human evolution, particularly in the course of brain expansion? It seems likely that the specific problems confronted by hominid populations kept shifting, which affected their most useful mental procedures. Social thinking and interdependence became important during habitat change. Indeed, the key social dilemma may have been over matters of environmental uncertainty, including the responses of others to novel circumstances.

The means for solving problems of a changing nature are distinctive to the human mind, reflecting both the social and ecological spheres. It is likely that these mental mechanisms evolved mainly over the past 1 million years, particularly as symbolic complexity infused many aspects of human life. This does not mean that the human brain is a general, all-purpose problem solver—a concept that is anathema to the evolutionary psychologists. Our findings are, however, at odds with the idea that the unique aspects of human reasoning evolved in a single, stable prehistoric environment.

4

Starting many years ago, clothed in different assumptions, I now find it impossible to deny that human evolution coincided with an awesome oscillation of habitats. *Homo sapiens* owes its existence, at least in part, to the uncertainty and depth of nature's perturbations. It seems inevitable that this founding basis of human life is manifested in the modern world.

The ways and means of environmental interaction unique to modern *Homo sapiens* reflect the ecological conditions of humankind's origin. The shifting, unforeshadowed settings of the Pleistocene favored faculties sensitive to environmental change and capable of stabilizing human needs. During hominid evolution, the mental, social, and ecological paradigms of our early ancestors were altered in ways that heightened the flexibility of response, the reading of environmental nuance, and the heterogeneity of behavior.

In the surviving species, these effects ultimately blossomed into symbolic coding, complex institutions, cultural diversity, technological innovation, human occupation of Earth's diverse biomes, an ability to recover from disturbance and grow by colonization, a greater awareness of self and of external factors, and the tendency to buffer environmental disruption by altering immediate surroundings.

A species whose evolutionary history has been spurred by environmental disturbance over long spans of time is an organism designed to ensure its ultimate survival against those shocks, to exploit stability as fully as it can, and to modify the risks of a capricious natural world whenever possible. Humankind expresses its ecological tendencies—its interactions with habitats—in ways that mirror the conditions of its evolutionary origin.

Our species' most potent characteristic is its ability to modify the immediate surroundings. The first results were modest: Put on hides for warmth and protection; share resources with others likely to reciprocate; make a garden to control food supply; find a way to heat and cool your home. Passed around the globe, these alterations have added up to a true metamorphosis. Our impulse for innovation probes the environment on an expanded scale. The result is a species now standing squarely in the arena of habitat alteration. By a process of origin responsive to nature's whims, our kind has become an agent of environmental change.

That our species reflects the ecological conditions of its ancestry is no mere metaphor. We have deduced, in fact, a process by which environmental disparity has spurred the origin of our strongest defining characteristics. Variability selection refines any tendency or method that enables populations of organisms to modulate the disruptive effects of environmental extremes. It places a premium on any tendency to try out novel systems of behavior as insurance rated against an unstable ecological past. Thus major *directional* advances, as in the organs and behaviors serving human intelligence, awareness, and technical mastery, may evolve uncoupled from any single state or directional trend in environment.

If this evolutionary process applies only to humans, however, the power of a general evolutionary explanation is lost, and we face the old dilemma, that in us nature has produced a freak of nature. This line of thinking is wrong for two reasons: First, the processes of evolution are manifold, and each has engendered its own unique models. Darwinian selection has built the greatest of all land speeds into the slender legs of the cheetah. Sexual selection once built a monument of antlers atop the skull of the male Irish elk. Kin selection has instilled the most exquisite social interdependencies in colonies of ants and bees. Variability selection has resulted in extraordinary faculties enabling human beings to manipulate the environment. In other words, the various operations of evolution have reaped wondrous and singular products.

Second, although the living human species seems to be stamped by the die of a mutable history, hominids were not the only large mammal so affected. Organisms exquisitely adapted to specific ways of life may shrug off certain rigid commitments of their kind and alter their ways of life in response to novel conditions. The success of these organisms and, ultimately, their species, as we saw in the mammals of East Africa, coincided with sharp reversals in environmental trend. I contend that the advantage these lineages gained over their more specialized relatives was a direct result of these oscillations. Globally, these were the most severe alterations in survival conditions in recent Earth history, and they were translated into local realms where they affected the evolutionary fate of animal lineages, some more than others.

The lineage most clearly disposed to this process was that of an upright sub-Saharan ape. Early on, with its bipedal peculiarities, it reconciled the conflict between habitats with many trees and habitats with none. The probings of the massive-toothed australopiths; the awkward handlings of lithic teeth; the trials of larger brains, some connected with large bodies and some with small; the perseverance of the australopith body design and its later discard by other lineages, all occurred in the tension between environmental decline and oscillation.

If, at that moment, global surroundings had permanently stabilized—if the savanna in which robust australopiths lived had become the irrevocable norm of nature, or if glacial sheets had expanded into the temperate zones and stayed that way, keeping world temperatures in the icebox zone—the bipedal apes and the dilemmas of the planet would have come to very different destinies. As representatives of the human experiment, we might now be highly successful gentlemen of a nutcracker lineage, or ladies of a thriving Neanderthal tribe. Humanity, we may imagine, would be matched to these environments, constant since some deep time in our ancestry.

But this is not what happened. The key developments in human origin began instead to mirror the disparities of Pleistocene surroundings. The constant handaxe was discarded. Diversity replaced sameness. Innovation took the throne next to imitation, while the hook of novelty helped to part the tethers of repetition. Where once there were none, imaginations and utterances of both great and transient meaning were added, now whisking around the globe as a new layer in the atmosphere. These distinctions of humankind are jewels set in the long history of environmental fluctuation.

I call a general statement of my thesis the principle of ecological heterogeneity: *The ecological characteristics of any species reflect the degree and pace of fluctuation in the factors pertinent to its existence during the period of its origin.*

This is not based strictly on human ancestry, but on the logic of what survival mechanisms would assist an organism and its genetic persistence through time in the face of environmental fluctuation. Scrutinizing the history external to human beings has an important impact on our view of human origin. We begin to detach ourselves from the traditional account of our place in nature, which is based entirely on the current appearance of the world.

The traditional account sees the inevitability of qualities distinctive to present human beings. At its best, it gives a gloss to the presumed superiority of the human species in nature. At its worst, it peers through cultural lenses that fill the world with social prejudices and rationalizes the biases of the immediate present as an outcome of an ancient evolutionary past.

Our approach to human ancestry differs by comparing innumerable clues of past environments and building a global history from them. We thus see that our species owes its origin and character to processes that have operated widely in nature. This ecological approach may begin to develop a more readily testable set of ideas about human evolution. Possibly it will be less susceptible to myths of superiority. I believe this to be a sign of maturity in the pursuit of our origin, and ultimately in our understanding of human existence.

As our investigation has proceeded, a number of possible histories might have been discovered. Since the time of the early Miocene apes, spans of environmental quiescence and chaos are both evident. There seems little basis, however, for denying that the irrepressible hominid *Homo sapiens* emerged in an especially mutable period of the natural world. In every well-examined chronicle, there are acute turns in the environmental course associated with human ancestry.

Still, there is something missing. We have so far gleaned only the story of an undomesticated world, without fences, rows of plants, or planned geometries. We have yet to grasp the most glaring development in the ecological genesis of humankind, the masterful new order of human-dominated ecosystems, oriented toward the production of food by agriculture and husbandry. With this shift in way of life, humanity has reconfigured the energy flow and diversity of Pleistocene habitats.

Is this most potent transformation in the recent history of the biosphere also a projection of the ancient human past? The birth and maturation of food-production ecosystems follow the ecological saga of human origin, mirroring the hominid ability to endure instability and buffer uncertainty. The qualities of agricultural ecosystems, which lie at the root of civilization and modern industrial society, echo the human interest in striving to avoid nature's promise of environmental disturbance.

5

Human control of food production began in various parts of the world about 10,000 years ago, and represented a dramatic culmination in the ability of human populations to buffer variability in environmental conditions.

The rise of agriculture in the Oaxaca Valley of Mexico at the cave site of Guilá Naquitz makes the point. A team led by Kent Flannery of the University of Michigan conducted a ten-year study to determine why the shift from food foraging to a settled way of life supported by the growing of crops took place. According to detailed pollen analysis, the environment of Oaxaca alternated between closed forest, open woodland, and mesquite grassland over the past 12,000 years. Vegetation zones shifted periodically, dissolving the stability of habitats that lasted for about 500 to 2,000 years at a time. First pine forest was dominant; then a mixture of oak and piñon; then a thorn forest associated with lesser pine, oak, and mesquite habitats; next an oak forest, and so on.

Despite this variability, the food remains preserved in the cave of Guilá Naquitz are remarkably consistent from one archeological layer to the next. In Flannery's words: "The macroscopic plant remains suggest that this area has undergone no *major* climatic change over the period since about 8000 B.C., although the pollen data are reported to show significant periodic fluctuations. . . . It will be clear . . . that there was a certain stability to the use of food resources at Guilá Naquitz. In general, the 15 most common plant and animal genera used in Zone B1 [about 6670 B.C.] were the same 15 used in Zone E [before 8750 B.C.]."

The foraging peoples of the Oaxaca Valley used two strategies in their early attempts to control food production. A model developed by Flannery and Robert Reynolds shows that early use of cultivated plants reduced the time people spent searching for food. In wet years, when food was plentiful, people sowed domesticated plants close to home. During dry or average years, they stuck to the old, conservative ways: They gathered piñon nuts, heart of agave, and acorns; they preyed on deer, rabbits, and turtles. Although it took more time, gathering and hunting compensated for the times when experiments in cultivation failed. Combining the two strategies reduced the differences in productivity between wet and relatively dry periods.

In hindsight we see cultivation and food-foraging as contrasts. But the Oaxacans apparently drew them together into a single way of life. According to Flannery, the beginning of agriculture in Mexico was "not a startling innovation or a response to some new stress, but the extension of a strategy already displayed by the group in preagricultural time." The sequence of food remains from Guilá Naquitz, one layer on top of another, reflects the way the people stabilized available foods by gradually adjusting the kind and frequency of plants they used. Early food production made life more secure despite shifts in climate and vegetation.

As the people of Oaxaca began to tend crops, they relied on certain enduring attributes of the human lineage. They buffered the disruptive cycles of nature that most threatened human security. It took farming several thousand years to be permanently adopted, but by the time this occurred, ways of storing cultivated plants and protecting domesticated herds had been devised. People extended the ancient art of transporting resources, brought the seeds of favored foods to their living sites, and expanded the sphere of the home base to include areas of cultivation and grazing. For the first time, there was a sense of ownership and control over fields and pastures. These novel behaviors were spin-offs of earlier evolved faculties that allowed humans to act effectively in the face of environmental disturbance.

Although ownership might have encouraged a sense of responsible conservation, much of the wild vegetation was depleted or removed as people planted crops and spent more time in one place. Even the oldest ecosystems of food production show a powerful link between human behaviors that buffer natural variation and provide security, on the one hand, and create intensive use and deterioration of particular tracts of land, on the other.

Environmental disturbance at the end of the Pleistocene, 10,000 years ago, was not unlike earlier periods. One of the best ocean records of this period comes from a core drilled off the northwest coast of Africa. When domestication of the land began, sea surface temperatures were undergoing a round of intense and rapid oscillation that lasted about two centuries. These oscillations occurred within a longer cycle when average ocean temperature fell 3°C in 750 years, followed by a rise of 5°C in 1,250 years.

What humans were then able to do that earlier hominids could not was apply an amazing compound of evolved capacities. Symbolic thought, mental creativity, imagination, complex cultural institutions, home-base behaviors, intricate social reciprocity, and long-distance exchange of resources were all now available. The annealing of these qualities during the late Paleolithic, dating to the time of the oldest cave paintings, foreshadowed the cultural development of food production and, later, the rise of cities and state societies.

Like the cheetah with its speedy legs, the bees with their complex societies, and the elk with his prominent antlers, humans had developed their own trademark, which was aimed at enhancing their own security in an uncertain world.

Control of food production was the most astonishing of all human capacities for buffering environmental disturbance. By tending the growth of certain species and modifying the landscape, people gained control over the flow of food. They could select and harvest plants and animals on the basis of their capacities to feed people. Finding food became secondary. Suddenly, the flow of energy within ecosystems depended on more than just nature's caprice. With the control of food production, human-dominated ecosystems were born. In an ecological sense, the modern world in which we now live had begun.

The changes humans have brought about in their new-styled ecosystems are a miniature portrait of the far longer environmental history of Earth. It should not surprise us that grasses and herbaceous species were the first to be selected by the earliest gardeners in various parts of the world; the spread of these plants, independent of human beings, was the most obvious change in vegetation over the past 50 million years. Human

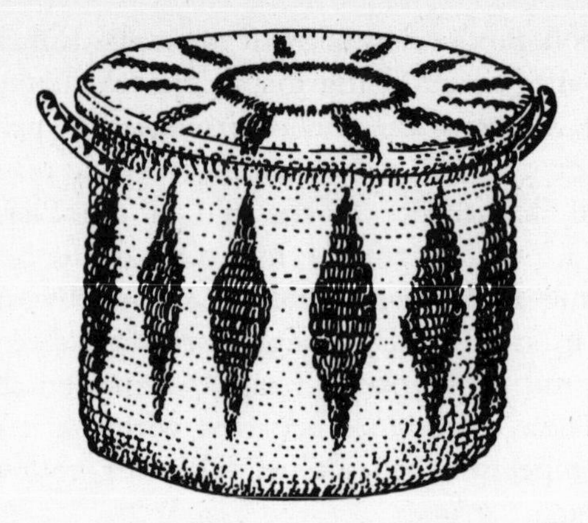

populations simply took advantage of the success of the wild predecessors of wheat, barley, rice, maize, and sorghum. It was the meeting of two separate pathways—animal and plant—in the ecological history of the Cenozoic.

It should also not surprise us that people favored seedlings that germinated quickly, grew rapidly, and remained dormant in unfavorable conditions—the very qualities that enabled grasses and herbs to become widespread in the first place and allowed plant populations to recover easily from environmental disturbance. In virtually every place, the earliest farmers had a choice between woody perennial plants and weedy herbaceous species. In every case, the weeds won. At Guilá Naquitz, the local plants highest in protein included susi nuts, guaje seeds, and mesquite—all woody shrubs and trees that take years to grow to maturity. These were not favored by the earliest farmers. Another species high in protein was the cucurbit or summer squash, an annual. Flannery describes cucurbits as "weedy camp followers," capable of doing well on disturbed soils. By favoring the cucurbit, the early farmers of Guilá Naquitz extended one of nature's primary themes. In the absence of that theme—adaptation to disturbance—neither the cultigens nor humankind itself would have been available to create the world that now prevails.

By definition we despise weeds. In certain parts of the world, more than 30 percent of agricultural productivity is lost to weed infestation. Yet these loathed plants owe much of their character to an evolved responsiveness, perhaps not wholly unlike our own. Cereals and other annuals we grow also have a weedlike resilience that parallels the adaptability of human beings. We met on equal terms, signed a partnership, and have thrived together while displacing much of the rest of the biotic world wherever we and our favored weeds happen to coincide.

During the Cenozoic Era, wooded grasslands and open prairies spread at the expense of forests. Most groups of land-based animals responded to this shift. But for humans, access to the dominant plants—the grasses and herbs—was indirect while they relied as best they could on scavenging and killing large mammals who ate the grass. The ability to cut and consume the meat of a gnu and an elephant's innards was the first turn in hominid diet, with tremendous consequences.

But consider a species well endowed to endure environmental disruption, which already possessed a diverse hunger, an innovative mental awareness, and the ability to make tools—and which then finds a way to consume grasses, directly linking it to an extraordinary and previously untouchable avenue of biological energy. This is a species that has finally tapped virtually every relevant aspect and trend in the environmental history of the planet, a species that has gone a long way toward making the planet its own. This event was the second major dietary turn in the human story—the consumption of starchy grains, growing them in massive amounts, using them to simplify and rechannel the productivity of the land, and creating a new landscape that serves the security of human populations.

The web of interacting animals and plants was reduced; the wide diversity of species was moved aside in favor of a smaller number. And so we come upon still another way in which the development of human-dominated ecosystems repeated an older theme of nature—the displacement and extinction of species. By manipulating the seed beds of the Fertile Crescent, Oaxaca, and elsewhere, humans joined the list of prominent factors disruptive to habitats. As farming intensified, humans became agents of the survival of the generalist, to which earlier hominids were subject. Controlling food production favored colonizers and resilient plant species. It marginalized the delicate and the specialized, which would otherwise fill Earth's bucket of biological diversity.

These same activities ultimately caused fields to displace forests and woods, recapitulating the long series of events before farming began. People today also replant trees, creating forests in open terrain once disturbed by agricultural activities. We think of this as good, but it is an expression of the same power that led people to destroy forests in the first place. Human populations repeat the kind of alteration that blazoned our heritage, eerily changing the landscape from treed to treeless to treed. We have assumed the power to reverse the conditions under which other organisms live.

Agricultural ecosystems have spread with alacrity, approaching the quickest habitat transformations of the Pleistocene. Human-dominated ecosystems thrive on a "movable feast"—the extraordinary exchange of food plants and animals across regions of the globe. Many food species show an adaptability to new places never previously encountered. The very qualities that allowed them to survive alterations of the Pleistocene make them hardy candidates for transworld travel and exuberant growth.

The symbiosis between these plants and their human caretakers moderates the effects of disturbance. Fields and domesticated pastures are protected, and we tend to think that we are doing the nurturing. But it is the yield of the fields and pastures that supports human populations, that allows simplified agricultural ecosystems to exist, that prompts a sense of human security and status quo against the fear of alteration. There seems to be no end to the insurance policies we need in a world prone to change. As human populations have grown, the image of what constitutes personal security has also had to enlarge. The human species now leans its ponderous weight on the productivity and control of food-production ecosystems, as on a crutch.

Agricultural ecosystems are strangely consistent with nature's ledger. Over the centuries, people have made fertile areas barren and barren areas fertile. *Homo sapiens* substitutes fields for trees, favors rapid reproducers over slow ones, promotes colonizing species, decreases overall diversity, instigates new ecological pathways that thrive on quick turnover. As humans exercise these potent responses to nature's own disruptive tendencies, our special ecosystems move like a tidal wave, unanticipated by organisms who are drowned or transformed by them. In the light of the thesis presented here, we should not be surprised by these clever reflections of Cenozoic history that lie at the root of human dominion.

At the same time, we cannot fail to be surprised by the implication: The collective power of humankind to modify environments, to create the most significant change in the biosphere since the descent of four-footed herbivores 250 million years ago, mirrors an era marked by remarkable strokes of habitat change. We are the hallmark of an age shaped by erratic cycles and throbs that altered Earth's settings in geologic moments measured in tens of millennia. The human species is neither a freak of nature nor separated from it.

Out of nature's own alteration, a hominid evolved with an armature sensitive to changing conditions, with the mind, hand, imagination, and social prowess to endure the best and the worst of times. In contrast to anthropology's prevailing story, the ecological character of this hominid was not born in a narrow ancestral habitat. This ecologically disruptive species, whose reflex is to channel energy inward from every accessible outer source, is a descendant of vicissitude who has withstood the hard prehistoric tests posed by broad variation in its surroundings.

Over the past 5 million years or so, a thousand-year stability was bracketed between strong tides of uncertainty. Fitful remodeling of habitats settled into a status quo that lasted for centuries. This in turn gave way to an erratic agenda, and so on. With the development of complex cultural mechanisms, the human lineage committed itself to the most profound means ever evolved of reacting to noise, fluctuation, fashion, and temporary advantage. Completion of the Monolith of Culture—the rise of complex symboling, innovation, and institutions—was itself an offspring of this heterogeneous era. As a consequence, our species possesses the tools for stockpiling strategies, beliefs, and technologies able to probe the biotic, geologic, and atmospheric domains of our planet. Human-dominated ecosystems display the propensities forged earlier in the long context of nature's alteration.

Our pivotal finding is that, rather than being against nature, our apparent dominion is based on actions that echo and amplify the kinds of natural alteration that led to the origin of humans in the first place. The supposed conflict between man and nature—the culture-bearing exploiter versus the original stable state—crumbles in the new grasp of how mankind evolved.

This strikes a blow at the body of belief that mankind's descent and uniqueness are separate from the natural world. It also denies the ill-begotten decree that nature exists for the purpose of human benefit or exploitation. Both views assign an illusory passivity to nature. Our origin is fully located in nature, connected with it in ways that we would not have anticipated.

Ecosystems manipulated and controlled by humans have prevailed for a brief second in the minute of hominid tenure on Earth. The span since mankind first sowed a seed or tended a lamb is long enough only to have experienced a minor fraction of the vacillations that have befallen the planet. Cities and state societies have stood for an even shorter instant. The modern age of expansion and pervasive environmental impact has met with the barest dash of nature's displacements. The industrial spread of humanity has occurred during a short quiescence—a time so brief as to permit human activity to outstrip natural tides of habitat change.

The question is, what happens next in the cross between the economic status quo, agendas to improve ourselves, and the natural world's altering tendencies stretched out unknown before us? This test, not yet taken, is one that will determine the future of the current experiment in being human.

CHAPTER VII

THE LITMUS TEST

I

IN the modern world it is commonplace to consider humanity a mere speck in the cosmic dimension of time. Evolution's conditional nature, which denies the inevitability of mankind's rise, is said to dissolve any notion of life's ultimate meaning. This philosophy of insignificance depends entirely on our lack of connection with something far larger that is truly significant.

If we grasp our evolved origin, what we find is not a trivialization of life or a lesser meaning, but a sense of life at a deeper level where our kind exists as part of a profound venture; where we are indelibly connected with all other living things; where we are as firmly embedded in the processes of nature as all other beings on the planet; and where humans now profoundly shape the present and future of Earth. This shared story and the continuity from one living thing to another have immense significance for the species that craves significance. They impart a sense of place in the turn of life. Our affirmation of this bond of blood and history is a measure of how far the philosophy of insignificance can be cast aside. Already we know that its basis—disjunction, separation from the natural realm—is invalid.

The word "indifferent" could be used to describe the dimension of nature in which humanity evolved. It is a frightful word, but it concedes the unexpected, wrenching events that people have had to endure over the course of history. That humanity has from time immemorial had to persevere in a world where such tragedy is possible need not disparage us or detract from the larger story. Rather, it draws attention to the remarkableness of our existence and dramatically colors the meaning of human resilience. That such a world has happened to preserve us as part of its rich living fabric is extraordinary, not a threat to our instinctive search for meaning but the very kernel of meaning. A world where our lineage, under trial, has made something of itself is perhaps to be more prized than one believed to consist of promises or guarantees.

Many of those writing about Earth's environment today think that scientific analysis erodes a deep and abiding mystery. Analytical knowledge squelches the wonder of nature and degrades what should be our soulful connection with it, which is revealed mainly by encountering it in the here and now, holistically and spiritually.

While respecting this perspective, I could not disagree more with its view of the scientific. By probing the natural world, we are seeking to grasp the primal connection we have with it. So far our pursuit has revealed myriad amazements that we would never otherwise have imagined: the role of planetary cycles and solar radiation in creating a dynamic place where life has evolved; the connection between our lineage and Earth's volatile history; the subtle and flagrant events—a minor drought, the uplift of large regions—that have touched all living species; the surprising parallels among distant organisms; the unique place we now occupy in a tumultuous period of Earth's environmental history; and finally, the connection of blood and kinship that we, as evolved beings, share with all other organisms.

These are things that could be disclosed only by seeing our deep past as something to investigate. The scientific motive animates our place in nature and uncovers a dimension unable to be revealed by experiencing only the present. Knowledge in this sense is where wonder begins, where a new sense of meaning, far deeper than earlier versions, may grab us and fill us with awe about the natural process of which we are a part.

The central idea of this book is not a theory of history; it does not seek to explain the particulars of recent human times. It is a theory of how the *range* of human possibilities known to us today could conceivably have emerged in a single, durable lineage. Theories of change almost always focus on directional change, perhaps because the overall trend offers an easy way to explain developments over time. Unexpectedly, we have discovered that direction results from widening the degree of variation. The Cenozoic became cooler and drier as global environments became more diverse over time and space. Human dominion and the innovations that made it possible emerged in response to extreme variability, rather than to directional change in habitat. What appears to be a progressive change in hominids was not a matter of uniform cause and effect, but a process of coping with an erratic context. The diversity and innovation of modern humankind is a consequence of this process, in which variance has played a larger role than the trend or average.

People everywhere are fascinated by extraordinary human achievements. Indeed, one of the questions most often asked by public audiences is how evolution could explain Einstein's intelligence or Beethoven's genius, Edison's insights or Solomon's uncommon wisdom.

I might add that singular accomplishments of the hands, feet, voice, and concentration are just as astonishing. They can be seen in the twirl of a great dancer, the passionate resonance of a diva, the diving catch of a wide receiver, the speed of an Olympic swimmer, the stillness of a monk, the artistry of a master carpenter. How did our evolutionary past make possible these peculiar powers? Many aspects of human life illustrate the capacity to break the mold of our expectations. What makes the feats of Einstein, Beethoven, Edison, and Solomon extraordinary is that they went beyond common thinking and doing. They broke the mold of how people thought and what people of their time did. Considering the many areas of human life, we find any number of people equipped to surprise us, expressing their human qualities in some novel way. Paradigms are made and broken; records are set and dashed to pieces.

We cannot comprehend such curious events if unique human qualities reflect a response to constant or stable factors in our deep ancestry. But if our physical, mental, and social origin came about by a lengthy process of surviving new contingencies that went beyond prior experience, they can be understood. The potency of human life resides partly in our evolved mental and social means of rising above the norm, the ability to say and do the unexpected. In this area, where previous evolutionary accounts have faltered, the idea of adaptation to variance and novelty makes sense of the diverse, unique, and surprising contributions people are able to make to human society.

The symbolic, meaning-seeking center of human life seems to reflect a dynamic, uncertain ancestral world. Consciousness is among the most precious human gifts dependent on the symbolic realm. Self-awareness is a sacred label, the flame we use to cast light on our innermost processes of thought, feeling, and wonder. The internal marvels of the self are penetrated by our passion for symbols, to name things we cannot see but can still sense. This is where we find the human soul—the reality of our deepest invisible resources of personal meaning, striving, and fulfillment. As we call upon these inner capacities, we exercise the ancient need to think and to search beyond our usual bounds. In this light, we are able to pursue the transcendent, a spiritual sense, which breaks the mold of the ordinary.

It is no less meaningful to live in a world ruled by contingency. Our instincts for symbolism, hope, and transcendence are dedicated to a world of uncertainty, full of surprise. Human beings have no choice but to be involved in its continuing creativity.

Our sense of connection with nature, of being part of something much larger, also has meaning for humanity's present dealings with the environment. Our origin, as part of nature, is a cosmology of inclusion rather than exclusion, a dramatic symbol of our involvement with Earth and its biota. Yet people all over the world are taught to believe the opposite, the value of human distinction and superiority over nature. Many people still endorse the idea of a separate and immutable creation, preferring not to accept, much less celebrate, our kinship, birth, and perseverance on a changing Earth. This difference is not a matter of abstract philosophy; it influences practical viewpoints and approaches to life. The view of nature we embrace will have a strong impact on the environmental dilemmas that lie ahead.

2

Disturbing tensions resonate today about our planet's environmental future. Thick walls divide us on the question of what that future should be. Who is right, the environmental doomsayer or the technological optimist? Is the backbone of nature about to fall apart because people are changing the world? Or will our technical know-how improve the land for humankind, keeping well ahead of any disaster? The conflict between these perspectives is daunting, and it is uncertain whether the opposing sides can ever be reconciled.

Surprisingly, such conflicting viewpoints have emerged from remarkably similar illusions about how the natural world operates. In one form or another, we seem to need the enduring myth of stability, even of a changeless natural order, with a special status reserved for mankind. Much in our individual lives and the status quo functioning of society are predicated on this ponderous creed.

My final aim is to illustrate how prevailing environmental viewpoints are underlain by a deep belief in a world that exists only in short-term vision, a world that is contradicted by the planet's archive of past habitats, a world where semblances of stability arise only if we ignore the longer corridors of time in which the human story has unfolded.

If the vital relationship we have uncovered is true, if the defining character of humanity—the social behaviors, environmental strategies, and universal institutions unique to *Homo sapiens*—developed largely because nature is prone to vary, then a predicament arises. If the idea of nature's original stable status is no longer viable, what about the powerful social ideals and economic systems nurtured by this myth? Can such systems of behavior and thinking still prevail? Or, since they are based on a mistaken creed, are they ultimately doomed?

A number of instances come to mind where human welfare has been seriously hurt by misconceptions about the fundamental workings of nature. Medieval physicians in the time of the Black Death assumed that all aspects of human health were determined by planetary positions. The lenses of astrology were blind to the real causes of the plague, which took an estimated one third of the European population during the mid-fourteenth century. Religious authorities, on the other hand, saw the pestilence as divine punishment, leading the Church to authorize penitential gatherings, which only furthered the spread of the fatal infection. The solution available at the time—halt the contagion by removing human bodies and feces from the streets and fields—was unable to penetrate the citadel of assumptions about how the world operates with humans as part of it.

A second example, from the twentieth century, concerns the dogma of plant genetics and growth set forth by the Soviet botanist T. D. Lysenko. Lysenko's fantastic precepts, which ran counter to what geneticists already knew, dominated thinking in the Soviet Union from

1935 to 1965. Lysenko believed that the productivity and genetic makeup of plants could easily be altered from year to year by manipulating their growing conditions. Sanctioned as official doctrine, Lysenko's ideas were translated into agricultural policy. The consequences were dire as crop rotations failed, soil nutrients were depleted, and acute food shortages arose. This illustrates the tragedy that may result if politics and laws are directed by an erroneous vision of nature's basic operations.

Tenacious ideas rooted in the soul of society, and ultimately in its institutions, have sometimes led humanity to difficult times. At stake here is whether our future is well served by an outlook on nature that may also be seriously mistaken. What will happen if our systems of behavior and thinking continue to be rooted in beguiling beliefs about the stability of nature and the separate status of mankind?

The feelings of doom currently hovering over us are, in my opinion, heightened by this false doctrine. The fiction of Earth's immutable natural state, altered only by human hands, will make harder our struggle to find the right thing to do. The deeply rooted creed of human separation from nature will continue to blur our perceptions of the environment and how we should proceed.

In the end, we must confront our most persistent myths. The challenge ahead is to uncover new ways of seeing our creative purpose and capturing the great collective energy of human life. Such new meanings will be nourished by coming to terms with our origin in nature and our permanent connection with the world that gave rise to us.

The two main outlooks in the environmental debate today are like two hostile cultures, oriented by different icons and meanings. I call them the Culture of Preservation and the Culture of Growth. While they appear to be opposites, their divergent views about the human-nature relationship are actually inspired by a common mythology, which is evident in the debate about our deepest environmental concerns today—global warming, population growth, species loss. Not every speaker on these topics fits neatly into one perspective or the other. Yet this kind of polarization claims center stage in the environmental debate today.

The Culture of Preservation: According to this culture, the world is divided into two conflicting realms, human and natural. The human realm is the catalyst of technological change and environmental disruption. Nature, on the other hand, is stable, and its original status is now deeply marred by human activities. This worldview, strongly evoked by the environmental movement, warns of nature's demise and of disasters arising because of human activities that oppose the rich but limited productivity and sensitive balance of natural habitats.

To many people concerned about the environment, maintaining a natural world untransformed by human hands is the greatest cause. An environmental ethic embraced by many authorities was originally inspired by conservationist Aldo Leopold's statement: "A thing is right when it tends to preserve the integrity, stability, and beauty of the biotic community. It is wrong when it tends otherwise."

In other words, stability is the original condition, and its preservation is the keystone of environmentally sound human action. Any measurable change from the status quo of nature means degradation of the environment. Such change comes about because human activity undermines the inherent balance of ecosystems and the economic benefit we and other organisms derive. Preserving the balance of nature is our moral charge, best accomplished by guarding against human excesses like overpopulation, air and water pollution, and the spread of people into natural habitats. Modern industrial activity is the main villain.

Global warming caused by industrial emissions and forest leveling has been an important focal point for the Culture of Preservation. Because of the greenhouse effect, which occurs as atmospheric gases trap the sun's warmth, industrial emission of carbon dioxide is held responsible for a 0.5°C global temperature rise during this century. Furthermore, it is generally agreed that industrial emissions will double the atmospheric CO_2 concentration within the next century, which is expected to increase global temperature by more than 2°C—an unprecedented rate of change leading to a world warmer than at any time in the past 200,000 years. Since trees absorb and store CO_2, our destruction of forests means that rises in atmospheric CO_2 will go unchecked. In the thinking of some advocates of preservation, global climate "will run out of control," and future humanity will be "living in a deteriorated world."

The dramatic rise in world population lies at the heart of the problem. Three hundred years ago the human population was approximately 680 million; fifty years ago, about 2 billion; today it nears 6 billion and is projected to reach somewhere between 9 and 12 billion in another half century. Doubling now takes place in a matter of decades, a rate of increase that is anathema to the Culture of Preservation. Natural habitats are rapidly taken for human purposes, and resources as basic as fresh air and water are exploited faster than they can replenish themselves. Given the population projections, the most resolute counselors of preservation fear that food supplies will be inadequate, leading to widespread famine. Human population increase will place an immense burden on other species, but nature's welfare is not the only thing at stake.

Much concern about environmental preservation focuses on the extinction of other organisms. The supposed balance of nature is girded by the diversity of species and ecosystems. Human intervention, however, is causing a calamity at least as great as the extinction of the dinosaurs 65 million years ago. The current extinction rate is extraordinarily fast and not limited to any one group of plants or animals. Biologist Peter Raven believes that one quarter of all tropical plants will be wiped out over the next three decades. Insects and other invertebrates responsible for the major flow of energy between plants and animals are expected to suffer tremendous losses—according to biologist E. O. Wilson, a minimum of 50,000 invertebrate species per year, due to the destruction of tropical forests alone.

The demise of particular species is not reversible, and any significant decrease in biotic diversity would take several million years to replace. To some, the protection of individual species is little more than "crisis management." The only long-lasting solution is to preserve entire habitats—species *and* the complex environments on which they depend. In this perspective the overall health of natural ecosystems is the critical factor, requiring properly managed land undisturbed by industry or agriculture.

The Culture of Preservation warns, however, that the tide of human activity is too great, making it impossible to preserve the land needed for the ecological health of the planet. Nature's balance is seriously threatened by population increase and climate change due to greenhouse warming. According to some predictions, society's response will not be adequate to maintain the ways of life familiar to us today. Coming generations may witness the disintegration of the ecological balance that has long supported Earth's biota and our civilization.

This dire message has been issued by many advocates of environmental preservation. In its extreme form, it considers humanity a blight, a viral strain spreading and killing nature. It calls upon human society to transform its activities and values radically before it is too late, to rethink its economic and political philosophies. To some advocates of this view, our current path will lead to "the end of civilization" and a planet "uninhabitable for people."

Sometimes, such worst-case scenarios are intended as a springboard for discussing solutions to environmental issues. A highly active contingent of people concerned about the environment remains firm in its call for a revolutionary shift in our ways. Its essential point is the need to alter the direction and pace of our economic activities. We must be prepared to lower our standard of living. We must drastically cut our consumption of fossil fuels because they emit CO_2 and contribute to climate change. We must use solar energy or find alternatives to oil, gas, coal, and wood. Industrial activities must be changed to cut the production of wastes that foul natural habitats. We must help nations in the tropics halt the process of deforestation, which destroys species and exacerbates global warming.

The basic message is that we must reverse the human tendency to disrupt the intrinsic equilibrium of nature. Our dealings with the environment require strict management. We must protect and enhance natural habitat reserves, which currently comprise less than 5 percent of the world's landscape. Where protection is not adequate, a plan for transplanting wild animals and plants will be necessary, since they will be unable to track rapid changes in climate zones during global warming. Laws must be enacted to limit industrial development and pollution. Strong incentives will be needed to help lower birthrates worldwide. The management of human affairs and our stewardship of nature must proceed hand in hand.

Detractors of the preservationist view portray such proposals as dangerously gloomy. It is unfair, however, to equate the idea of environmental preservation with its darkest statements. Many of its widely discussed hopes and goals are important for everyone to weigh seriously within the framework of existing creeds and ways of life. To many people, these goals are *right,* connected with long-standing beliefs about the natural world and the incompatibility between human behavior and nature.

The Culture of Growth: Opposed to the preservationists are many who see Earth largely as a stage for human progress. Instead of saving the rain forest and the spotted owl, or warning of the "end of nature," followers of the Culture of Growth promote modern technology, wealth, and personal economic freedom. They believe that the scale of our economic activities should increase. While environmental issues are important, human technological and social solutions will be able to meet whatever problems may stand in the way of progress. Our economic and technological activities are capable of offsetting global warming, pollution, food shortage, and habitat change. Growth proponents seriously doubt that we are in the midst of a global environmental crisis threatening humanity or any other part of life on Earth. They call environmental change a "nuisance" rather than a crisis.

Global warming offers a good illustration of the Culture of Growth perspective. Its views on warming are buttressed by several lines of defense. First, growth proponents question whether even a significant temperature rise has occurred in recent decades. Any warming probably reflects a natural recovery from the cold period known as the Little Ice Age between A.D. 1400 and 1890. According to journalist Ronald Bailey's book *Eco-scam,* this "natural recovery would wipe out the warming attributed to an enhanced greenhouse effect." The impact of human industrial activities is therefore minimized.

A different line of defense accepts that CO_2 concentration will double in the next century but believes that this will yield only positive effects. Since carbon dioxide and warm weather encourage plant growth, world food supplies will be enhanced. Desert areas may become fertile, as they did during a warmer Earth climate nearly 8,000 years ago. And, in a curious line of thought, it is predicted that the factors supposedly involved in global warming will actually promote cooling; greater water evaporation in warmer weather will increase cloud formation and so result eventually in cooler weather.

The bottom line of the growth perspective is its faith in technology and the advance of human know-how. Technology will enable us to adapt far faster than climatic change can possibly occur. Better plant varieties will be developed far more quickly than any projected warming due to industrial emissions. New pesticides, chemical fertilizers, and herbicides will be invented, and genetic engineering will produce high-yield crops to meet the growing human demand. These practices will prove far cheaper and less problematic than efforts to restructure modern society and economic values, called for by ardent preservationists.

In contrast with the preservation position—in which a doubled world populace, a proportional rise in consumption, pollutants, and other environmental assaults is "not only virtually unimaginable, but impossible"—the Culture of Growth welcomes the tremendous increase in the number of people, who will add their innovative solutions to the difficulties humanity may face, including new ways of overcoming nature's limitations.

As stated by politician Dixy Lee Ray, who champions the growth perspective: "What nature does not supply, man must provide. Improved varieties of food plants grow faster, become larger, and produce more harvest than the wild types from which they were developed." Human energies and technologies are not only remodeling the natural world; they *should* remodel it, for nature is something that serves humanity.

Ronald Bailey echoes this outlook: "Unlike most other species, we modify the world to suit our needs; we don't have to adapt to its given constraints. Human intelligence usually breaks the bonds that 'carrying capacity' imposes on other species. This is why economics, the science of human interaction, is more relevant to the study of human beings than is ecology."

These statements illustrate a vital dimension in the Culture of Growth—human separation from nature and superiority over it.

While rejecting the view that human beings are an integral part of nature, the growth perspective appears to accept the idea of nature's variability. According to D. L. Ray, "change is nature's rule. The entire history of Earth is one of climatic change." But this view is asserted only to downplay the significance of environmental change due to human activities. The advocates of growth portray climatic fluctuations as mild, involving departures and recoveries around an optimum.

Given the major advances in human welfare and control over nature during the past several centuries, environmental change is deemed largely unimportant to human affairs. Life span has been prolonged significantly; famine has been reduced to nil except where war has hindered the distribution of food; and the standard of living has risen for most people. According to the growth perspective, these trends will continue, spurred by human technology and a market economy. In short, the basic philosophy is that humans are unaffected by environmental change because of our ability to compensate when nature falls short in providing for human needs.

On the issue of biodiversity, partisans of the growth perspective have rather little to say. According to D. L. Ray's highly touted book, *Environmental Overkill,* we are not running out of species. And even if some species are endangered, the issue is not ecological but economic. Conservation is simply too costly to justify. The cost, moreover, is borne by taxpayers and by endangered industries such as timber and mining, which serve our rising numbers and modern lifeways. "This land is *our* land," according to this viewpoint—not land the government should control for the benefit of other species over our own.

A centerpiece in the environmental debate is the economic freedom of the individual. Reducing government costs and decreasing bureaucratic control over our interactions with the environment are thought to be paramount to human progress. International agreements that stipulate environmental policy are accused of undermining our economic rights. In the United States at least, these issues echo the politics of our time and tensions deeply rooted in American life. Conflicts between the two cultures mirror long-standing tensions in American history—government regulation versus individual rights, international involvement versus internal national interests.

The Culture of Growth argues consistently against international treaties and environmental expenditures required by law, because these take money away from industry and individuals. This contrasts sharply with the preservationist view that human effects on the environment must come under national and intergovernmental control; bureaucratic management is essential, including rechanneling wealth from industrial nations to the developing world.

The Culture of Growth focuses on advances in the standard of living and health of the general populace. The Culture of Preservation, on the other hand, believes the emphasis on industry and "progress" is retrogressive and dangerous. It urges the need to plan the health of ecosystems supporting all forms of life, and the control of human activities. Opponents revile this as a "back to the caveman" attitude, an ignorant and oppressive approach.

It is obviously painful to try to reconcile these two perspectives. The situation is not helped by the fact that exactly the same environmental data are often taken to opposite conclusions. For example, the ozone layer in the atmosphere has captured the attention of many environmentalists. Its measurable depletion over the past decade or so, thought to be caused by human emission of chemicals known as CFCs, permits more ultraviolet light to reach Earth's surface. The effect on crop plants has been investigated experimentally by botanist Alan Teramura. The preservationists note correctly that the majority of crop plants he tested were adversely affected. They consider this to mean that crop yields will drop significantly, corresponding to large economic losses, as the ozone layer becomes thinner. Therefore, human activities that cause the release of CFCs should be limited by law.

The proponents of growth see something completely different in the experimental data. More than 40 percent of the tested crops were unaffected by extreme ultraviolet increases; thus, even in the worst scenarios of ozone depletion, crop productivity can be maintained by selecting the right plant varieties. In fact, certain cultivars even increased their yields in higher ultraviolet radiation, as if to suggest that human effects on the ozone layer could only *improve* our situation.

Each culture sees the world through a distinctive lens. The Culture of Preservation fears the end of nature, the unraveling of the ecological web underwriting the weal of all organisms, including *Homo sapiens.* The Culture of Growth abhors the end of progress, especially demands for a lower standard of living in the industrial nations. Both fear the end of what they consider to be the status quo. For the Culture of Preservation, the status quo is the original condition, the health and equilibrium of natural habitats. For the Culture of Growth, it is the steady improvement in our control over nature, which has secured certain economic advantages. Humankind and nature stand as opposites—a distinction that pervades the key assumptions of both outlooks.

The dissonance between active humanity and passive nature seems fundamental, if not comforting, to many people. While declaring humans a product of nature, the preservationist agenda depends on an antagonistic separation and mismatch between human behavior and nature's way. Yet this doctrine is invalidated by the record of Earth's volatile past and our own origin in it.

The Culture of Growth errs equally by drawing the division between dynamic, inventive, problem-solving humankind and our relatively passive, malleable surroundings. This presumed state of affairs is what allows us to alter nature for human good. It permits human mastery over the environment. While nature may provide problems, humans eventually solve them. Human superiority stems from this illusory divide into natural and human domains. This divide has, however, never existed over the long course of human presence on Earth.

Why has this belief in nature's constancy, human separation, and superiority come about? Why does it continue to chain the way we think, even across adversarial points of view? Three reasons come to mind.

First, we consider our surroundings in time spans that focus less on nature's true workings than on short-term advantages and desires. We begin an irrigation project without thinking that rivers change direction or dry up. Since such changes are fairly rare, they are easily ignored, while personal gain is measured much more briefly. Yet we remain blind to the fact that rare events do occur and that long cycles of change are in the process of being effected. We build architectural wonders on faults or over water, disregarding environmental dynamics over the long haul. Pinned as we are to the presumed security of short time frames, we can hardly grasp the longer course and the larger outcomes.

The environmental debate is conducted largely in this narrow perspective. Dire predictions of environmental collapse are based on trends no more than a few decades long. As the critics of preservation point out, the threat of an ice age caused grave concern just two decades ago, while too much warming now raises the alarm. Both sides are preoccupied with the short term. Attempting to negate the claim that world mineral resources are overexploited, Ronald Bailey declared: "Not quite. The U.S. Bureau of Mines estimates that at 1990 rates of production, world reserves of gold will last 24 years, mercury 40 years, tin 28 years, zinc 40 years, copper 65 years, and lead 35 years." It is hard to believe that this is intended to be an optimistic statement. Yet in the blinding belief that Earth exists as a backdrop for human progress, it is a positive response: Other riches will be found for us to mine.

In the fiscal-year marketplace of debate, even three- or five-year trends are considered good currency. The point is that the time scale deemed relevant to environmental issues, on which we base our measure of security or fear, is far too brief to encompass really important changes in the global landscape.

A second reason why nature's stability has been accepted is that the environment may have been fairly steady in recent millennia. A study of one of the ice cores drilled from central Greenland points to erratic and fast switches in global temperature and ice volume over the past 150,000 years—with the *exception* of the last 8,000. Although earlier stable periods occurred—and were broken—the recent one is unusually quiescent compared to the unsettled time just before it. Although rapid shifts in habitat, not due to human cause, occurred in certain places during this span, the Greenland core suggests that climate overall has been less variable.

The past 8,000 years were also the time when prevailing beliefs about mankind's relationship with the world were developing. Could the mythic liquor of nature's original harmony—and our dissonant intrusion—have fermented in the social conscience because of the unusual calm? This powerful creed may reflect a real disjunction between human behavior and nature's expression over recent millennia.

A third reason ties even more directly to our main thesis. Stability and dominion may channel human thought because the ways we mitigate uncertainty and create security are so amazingly effective. We may cherish stability as the natural state *not* because nature is that way, *but* because the social milieu we create for ourselves is indeed relatively stable. That is, we see the world through lenses curved to match the intended state of our social universe. We project into our visions of nature the way we expect our relationship with it to be. With society as the grand leveler, we expect our natural surroundings to be steady and secure.

The mandate to mitigate change, to create personal security, is the luster on our profound adaptation to nature's surprises over the past 1 million years. Environmental uncertainties are buffered by our most defining social and psychological methods of survival. Thus we nurture in ourselves a distinction from and dominion over nature. This means, however, that a deep-seated tension exists between the glasses we make for ourselves and the longer external reality.

The influence of this impressive ideology—stability, separation, and superiority—is only as strong as the symbols that guide our belief in it. But here we meet with doctrines passed down through the centuries, powerful icons that still govern the way we think the world works. The independent origin and eternal separation of ourselves from all other living things is surely among the most tenacious beliefs in the history of human thought. And if nature and her offspring are immutable, then kinship among species is denied. Rejecting any genealogy among living forms is the vital prelude to the assumption of human dominion, for in the telling of any such story people reserve for themselves a special and independent place. This is seen in many origin stories over the world, from ancient Persia to the Pacific islands.

In the words of the nineteenth-century naturalist George Perkins Marsh: "Of all organic beings, man alone is to be regarded a destructive power . . . though living in physical nature, he is not of her . . . he is of more exalted parentage, and belongs to a higher order of existences than those born of her womb and submissive to her dictates." On this basis, Marsh implored people to halt their disruption of nature's harmony.

The same creed serves the advocates of economic growth equally well. Short-term advantage can easily be mistaken for long-term progress in a world where nature is steady and humanity stands above all other things. "We can effect positive change as long as we accept our humanness and know that we are a species apart from the rest of nature and beyond simple naturalness." Thus Dixy Lee Ray ends her argument favoring modern industrial progress while casting a dim view on present environmental concerns.

My point is not to undermine the importance of natural conservation, nor to question the pursuit of human welfare. My concern is whether these ideals can be framed to reflect our real involvement with natural systems. The extended perspective on time adopted here may be thought to be irrelevant to current environmental concerns. After all, the remodeling of landscapes over millennia hardly seems pertinent to the political process, economic strategies, and ecological hazards that currently occupy us. This, I would expect, is the majority opinion.

The alternative is to extend our sense of time relevant to human life, and to embrace the reasons concerning our existence as a phenomenon of nature. While seeming a little too abstract at first, this perspective actually makes very clear the discordance between our current short-term measures of gain and the long-term resilience that underwrites human survival.

The competing worldviews of growth and preservation are both deeply rooted in a series of questionable beliefs about nature and our involvement with it. Both cultures ignore the extent to which habitats are subject to complex forces of change over short and long time frames. And, ultimately, they veil the ways in which humans are now engaged in this natural dynamic.

3

All of our understanding about the environment ultimately passes through the potent force field of human value systems—socioeconomic, religious, political, and moral. All viewpoints and decisions we make reflect human aims in the end. This includes the strictest conservation ethic, and also the tightest defense of individual freedom to extract the resources around us. I've drawn the preservation/growth contrast to show the range of values people aver, and to illustrate the difficulty of coming up with a single philosophy about the environment and an acceptable plan of action.

If we are to believe either outlook, a lot is at stake. The very foundation of life on Earth is crumbling, and we are responsible, according to one view. The other believes that the factors—industrial technology, economic growth, rising population—deemed destructive by the first outlook are, in fact, a panacea. If we limit those factors, we undermine the very basis of human progress.

Are we about to enter an age of environmental misery? Are we to dread the extinction of a reckless civilization? Or should we, immensely capable of controlling our surroundings, embrace a technological and economic renaissance that ensures human welfare?

From all we've learned about Earth's deep past it may seem that the first serious blow must be struck at the Culture of Preservation. If, in time, nature abhors a changeless world, the change contributed by people is perhaps part of the way things are. If our ancestry took place under eventful, moody circumstances, it might be argued that our activities are consistent with nature's past insults. Nature is resilient, and we, as a survivor of extremes, are merely testing the boundaries, not exceeding them.

But this argument, which roughly echoes the rationale for ambitious growth in the modern world, is seriously flawed. The environmental debate has parsed the question of habitat change into pieces—acid rain, greenhouse emissions, the ozone hole, species loss, and so on. Arguments proceed on each isolated issue: How much change has occurred? Has it been significant? Have human commerce and industry been responsible? Often missed in the fray is the sweeping repainting of Earth's environmental canvas caused by the rise of human-dominated ecosystems. The spread of these simplified habitats has transformed the channels of energy flow on land and added ourselves to the vibrant causes of habitat uncertainty.

The opponents argue tediously about the amount, cause, and impact of environmental change. Cutting through the thick skin of the debate, we come to the heart of the matter. While our species consumes less than 1 percent of the primary productivity on land—that is, the total plant material available to all organisms—it dominates, alters, or destroys nearly 40 percent. Although human-dominated ecosystems are consistent with the environmental history of our era of origin, their impact overrides all other considerations concerning human influence on Earth's environmental fate.

There is no question as to whether human impact is sufficient to engage the spiraling trajectories of dramatic natural change. Human influence is sufficient and engages the causes and effects of past fluctuations. There is no debate about whether *Homo sapiens* alters the atmosphere, shifts the pattern of rainfall and temperature, rearranges the mosaic of food, water, and growing conditions, and culls those organisms unable to cope with the pace and pattern of change. These impacts are all fundamental to human-dominated ecosystems today.

Earth's climate system is very labile. Habitats on a local to global scale maintain a certain look for a certain length of time but ultimately move, reshaping themselves according to a complex union of causes: changes in solar radiation, in volcanic aerosols, in the planet's relief, in its plumbing and heating systems, in soils and the responses of organisms. In the past, massive belches of volcanic particles caused unforeseen secondary effects. The magnitude of these effects depended on incidental factors—the planet's orbital position, the pattern of ocean currents, the height of plateaus affecting jet streams and air moisture, and the amount of snow, desert, grass, and forest covering Earth's surface at the time. The orchestration of these factors led eventually to multiple and sometimes sharp environmental transformations, changing the habitats on which populations of organisms, including earlier hominids, had come to rely.

This union—that strange symphony of causes known as *nonlinear interactions*—still determines the environments of the planet. By any guide we know, these factors are ongoing: Climatic change will occur, and the terrain of plants and animals and water and soils will be remodeled. The real issue is not whether people are altering an environmental system that is solid, predictable, and stable. The natural world that generated the human enterprise has none of these characteristics.

Modern humanity alters the spatial arrangement of water and soils, the distribution of fields and forests. We explode clouds of soot and other particles into the air. We modify the chemical composition of the atmosphere, including gas concentrations linked with vast climatic shifts well before humans were a factor. We constrict and expand the range of particular plant and animal species.

In other words, we have insinuated ourselves in the old alliance of change. Once controlled by the ancient whims of habitat, *Homo sapiens* now brushes against the strings that govern those whims. Some of the strings, such as soils and vegetation, we pull with obvious goals in mind. But others, such as atmospheric emissions and changes in water distribution, are usually pulled as side effects of our actual aims. In the process we alter several variables that, at least in the past, were involved in cascades of environmental change. Humanity is the new contingency in the alliance.

The real problem is not that humans are anti-nature but that they are a sharp reflection of nature, a point largely if not entirely missed in current attitudes about the environment. Not by mythic ascendance but by difficult origin, the human lineage is now fully tapped into the fundamental signal of Earth's recent environmental drama. Evolved in unsettled times, the surviving hominid has entered the crossfire of causes that bend nature's course.

By redirecting the circuits of energy and simplifying the surrounding habitats, humans have instigated new and profound uncertainties. Our activities tend either to exaggerate or to dampen the effects of other environmental factors. In other words, we are an integral part of the nonlinear interaction among natural variables.

Consider, for example, the current rise in carbon dioxide concentration in the atmosphere. As in any other warm interglacial period, the atmospheric concentration of CO_2 is higher today than it was during a cold era. Burning fossil fuels and deforestation magnify this condition by transferring CO_2 from the ground to the air.

Imagine for a moment that this human contribution had coincided with glacial conditions. A dampening effect might have occurred, negating the glacial drop in CO_2 and keeping global temperatures on an even keel. But that is not what is happening; the amplifying effect caused by human activities further increases the concentration of CO_2. Scientists have developed models to show that the concentration will soon be doubled and global temperature will rise higher and more quickly than in any period of the turbulent Pleistocene. The exactness of this prediction hardly matters; the models, by design, are not supposed to factor in the enormous number of influences that cause climatic uncertainty. Because there is no precedent for the added human effect, it is impossible to know how large a change will actually take place. This promise of uncertainty lies at the crux of human life, as it did even before our activities reached their current scale.

In tilling the soil, choosing crops, protecting fields, and moving species around the world, human activities have favored organisms able to bend to novel conditions. Weedy plants, small rodents, and the invincible cockroach have been especially successful in adapting to nearly every human venue. But what about other organisms that survived the Pleistocene largely because of their ability to move, to blaze new paths in pursuit of vagrant forage?

When land is converted by farming, urban settlement, or industrial growth, people try to keep it that way. We displace other animals and plants and, at best, hope that well-defined natural reserves will serve as an adequate home for wildlife. With even minor shifts in rainfall and a growing population, the liquid complex of industry, settlement, and farming will spill into new zones to meet human needs. These are neither controversial projections nor matters on which the two cultures disagree. With this expansion, the main response evolved by plant and animal populations to deal with climatic change is rendered ineffective. If we combine species confinement, human expansion, and environmental change, all lessons from the past suggest that there will be a precipitous fall in the number of animal and plant species, beyond that currently taking place.

Alteration of the landscape will affect atmospheric and water circulation on a global scale, as it did in the past. Because human activities are embedded in nature's complexity, there is no easy formula for calculating our influence on nonlinear patterns of habitat change. Our recent cutting of tropical forests, when they have been relatively widespread, has not created a stable average. Instead, the result mimics the pattern of forest retreat during past dry periods. Inevitably, an arid phase will come that would normally shrink tropical forests to small areas when people have already done so. If our grasp of environmental history is correct, this dual impact will have large secondary effects, causing the reflection of more solar radiation from Earth's surface and weakening tropical forest involvement in the carbon cycle. These results will be compounded by the panoply of other environmental variables operating at the time.

This is the kind of intimate link that now connects humans with the forces of nonlinear change. Adversaries in the environmental debate all want to find a barometer to measure such change. They seek an indicator of how gradual rises in carbon dioxide or gradual falls in precipitation will translate into future climate. They expect to show how small changes in ozone or incremental losses of species will ultimately prove catastrophic, or simply unimportant.

There is no such barometer relevant to big ecological shifts. Environmental remodelings of the past suggest that large changes occur by thresholds, by the circumstantial union of many factors. The change is registered not like a barometer but like a litmus test in which no indication is noted—the litmus paper stays the same color—until a certain point is reached. At that point, the change means that a significant switch in the environmental mode has already occurred.

I agree with an increasing number of scientists who admit that a firm evaluation of environmental change, such as global warming, will be possible only when our measures of change begin to exceed known levels of natural variation. This, however, is a litmus test; when we finally see an adequate demonstration of the warming—and thus know its cause—the critical transition will already have taken place.

That it is so hard to discern human from other forces of change is a vital point. We are a cause of uncertain environmental change *not* because we are outside nature, upsetting its balance, but because human practices mesh so effectively with the long-standing dynamics of the world.

I am troubled, therefore, by one of our most common distinctions—human versus nature—and by the assumption that we alone control and disturb a passive natural realm. While many people realize that such beliefs are not really true, the most durable institutions in Western society were born in the light of this worldview. That it fuels the basic assumptions of both sides in the environmental debate is merely a reflection of its broader influence. Our economies, world political relationships, and the expectations of daily life are propped against this powerful backdrop.

Staunch advocates of growth say that we can count on vast technical improvements in the way we live off the land. Plenty of uncultivated land, they say, can be made to support the rising world population. This sense of improvement is locked into the key assumption that human technical knowledge is the only important variable, while external factors remain relatively constant. A reliable, secure environment is something we have always assumed will support our national and personal efforts of economic growth.

The global status quo—an interdependence involving resources, money, political power, and way of life—has been founded on a narrow path of world climate and resource distributions. Our world is committed to technologies, modes of commerce, and worldwide networks in line with the present state of things. In places less stable environmentally, we find peoples who were once less committed to market commerce and industry, but who over the past century have applied these approaches and now find themselves webbed into international aid and global policies built on short-term expectations, the current status of resources, and the permanent advantages of the powerful.

I do not know whether *any* past civilization has ever favored sensitivity to change over a belief in the fixed superiority of its ways. The news and anthropological writings are rife with portraits of perilous decisions people have made when they remained unmoved by context, numbed by traditional thinking. The question is whether our institutions, philosophies, and assumptions can assimilate what can now be learned about nature's altering tendency and our impatient hand in this realm.

If we have gained any insight in pursuing human origin, the word *contingency* comes to the fore. Our presence in the nexus of Earth's history is predicated on successful response to new conditions. If our thesis is correct, all that we can blame on our ancient past is its prejudice in engaging certain hominids to endure the novel, jarring bends in the evolutionary journey.

This view of human origin contrasts with all previous theories, which restrict humanity to a narrower destiny. They trace our ancestry back to a limited ancient habitat, to a certain way of life, and to a constraining course of natural selection. Consequently, there is no recourse but to lay blame on the narrow prescripts of that ancient era, or to rebuke our modern ways because they deviate from the more idyllic of those ancient charges.

In presenting his hunting hypothesis, Robert Ardrey has no alternative but to imagine modern humanity as an innately aggressive hunter. In discussing the idea of the "human zoo," zoologist Desmond Morris sees no escape from the inborn template of ancient tribal life, even in the urban cages where much of humanity now paces. And in his exceptional work *On Human Nature,* Edward O. Wilson claims that humankind possesses specific genetic adaptations to the bygone era of Ice Age hunting and gathering. In the modern era we must follow the constraints of "archaic behavioral adaptations," making the best of old adaptive tools honed in an ancestral environment now obsolete. Humanity's challenge is to become aware of the narrow course carved out in prehistory and to act accordingly.

Much in the functioning of human society has its roots in the deep past and is shared with other organisms in our common ancestry. The issue that concerns us is what lies on top of this. In what ways are humans unusual? What does our peculiar interaction with the natural world mean? The principle advanced here—human adaptation to novel contingencies—has a liberating quality. It means that our unique social behaviors, mental capacities, and ecological strategies are not attached to strict limits defined in our past. Rather, they arise from a series of programs and potentials sensitive to change, responsive to things happening around us. This system of behaviors is underlain by a genetic blueprint that opens up human possibilities, giving rise to people primed to discover what is happening in their surroundings. It inclines people toward solving challenges, as they are perceived, even if this means creating and absorbing new lines of information never imagined by earlier generations.

To be sure, the environments we inhabit today differ vastly from early human settings. Planted fields were not experienced by *Homo habilis.* Nor were cities occupied by *Homo erectus.* Indeed, for much of our evolutionary history, small hominid groups collected vegetables and meat with tool-assisted hands. They had no acquaintance with parades, traffic jams, or airline food. In no sense were these predecessors *prepared* to live in the habitats of our making.

But the wild savanna and the foraging life are not what gave the human lineage its peculiar trajectory. What did so was the evolution of social, mental, and ecological means of sensing the heterogeneity of surrounding conditions. The strange buoyancy of the hominids is in us, a hopeful heritage of response to novel environmental dilemmas.

Humanity now has a new ecology oriented toward controlling certain aspects of its surroundings. We are inordinate in changing the land we inhabit, wherever we live, which means change worldwide. But we cannot ascribe this state of affairs to some desideratum of prehistory, a scourge planted deep in our genes by millions of years of natural selection. Nor is it a violation of the supposed narrow human nature thus instilled. Our genetic blueprint enables our brains and societies to live creatively in an uncertain world. As the ability to use language is ingrained in us, so, too, is our sensitivity to novel conditions, without specifying the actual response.

This does not mean that older forms of emotion, self-centered reasoning, and group bias are turned off. It does mean that human societies have other possibilities. The term *human nature* typically connotes something inevitable about our behavior and destiny. If anything inevitable has been added to the long evolved repertoire of humanity, it is our sensitivity to the exigencies around us.

The future is not easy to foretell because the world's manifold affairs guarantee the unexpected turn. By virtue of an evolutionary origin, we are endowed with the means to respond, to work with the conditions at hand, and to imagine what to do next. We also now possess an *awareness* of Earth's splendid dynamics, its unending causes and consequences. In this lies our defiance of a dooming fate, for it implies that we possess ways for altering our transactions with the environment.

We can be sure that the multitude of factors causing environmental change will assert themselves. The result, based on past instances, is likely to foster change in the resources underpinning the current global state of human ecosystems. In other words, the reality on which present expectations and advantages are based is temporary. If our investigation has cast a distinctive light on the character of mankind, we can see that narrow, extravagant commitments are eventually buried in the dust. Shortsighted benefits do not add up; in time they are subordinated by the necessity of resilience. Cultural institutions that lead people to assume their secure command of nature for personal gain will eventually find extinction's fatal edge. Our long-term welfare resides instead in inspired tolerance. An ability to tolerate environmental insult is the hallmark of the human lineage; and understanding our origin has the power to inspire in us the principles of human existence.

At the heart of the matter is what we believe our relationship with nature to be. This area of belief, which defines the place of human beings in the world, envelops all societies with invisible, mythic tendrils. Do we hold a sacred place in creation ordained by separate origin and inevitable advance? We must now face the fact that we are not superior in these ways. As sweet as our ownership and as proud as our domination of Earth may seem, our existence as a species is a bitter miracle.

So we may find a different but still extraordinary meaning in our story. We are part of nature's ongoing process; it is no exaggeration to say that every person holds the seeds of eternity—life's primordial code—in each cell, and a range of destinies in our ecological convictions. *Homo sapiens* has come into being not by preordained progress, but by trial and delicate survival. Aware of this origin, we may now begin to see our dependence on Earth's processes and to realize that our influence leads us into the most delicate challenges and responsibilities. As we touch the factors that fueled even our own appearance on Earth, much depends on what we will now do.

Humanity exists not only in the present; we are a species through time. If the story of our origin has any further meaning, the sense of time beyond personal experience, beyond the desires and aims of a generation, will determine a great deal about the long-term future. To absorb this longer sense of time and its ultimate practicality, and to build it into what we believe about the human presence on Earth, remain a daunting challenge.

Humanity is also a global concept. We are connected worldwide by information, by currency, and by sight. Products, outlooks, and even entire institutions have permeated rapidly across cultural barriers. Universals woven from the many threads of human diversity are increasingly apparent. Without question, our lives will continue to be shaped in this larger setting. Our institutions will need to be responsive to the manifold contexts and nuances of what it means to be a human being. Without this diverse perspective, institutions that now provide for us will struggle, rutted in archaic values that do not apply in a changing world.

The outcome will surely be played out in world politics and economic policies. While this vast arena sometimes seems beyond our influence, fundamental personal decisions will underlie our next move, the acceptance of any plan, and the path ahead. We usually see our choices in terms of things to buy, or how to raise enough money to start a business or go to college. The matter before us is deeper. People like you and me will decide a course for the planet by choosing ways to interact with it. We cannot avoid being involved in the collective result, which, in a primal sense, is the continuing basis of human origin.

NOTES

PAGE

I: ORIGIN

5 Research at the Olorgesailie prehistoric site was first carried out by Louis and Mary Leakey, and by Glynn Isaac. See Isaac (1977), *Olorgesailie,* University of Chicago Press, Chicago. The Smithsonian and National Museums of Kenya research, conducted since 1986, is presented in R. Potts (1994), *Journal of Human Evolution,* vol.27, pp.7–24; (1989), *Journal of Human Evolution,* vol.18, pp.269–276; and is reviewed by Ann Gibbons (1990), *Science,* vol.247, pp.1407–1409.

6 The twenty-four-hour time analogy is in Kathryn Lasky (1990), *Traces of Life: The Origins of Humankind,* Morrow Junior Books, New York.

7 Charles Darwin (1871), *The Descent of Man,* Modern Library, New York; Thomas Henry Huxley (1863), *Evidence as to Man's Place in Nature,* Williams and Norgate, London.

9 Darwin (1871); also see Matt Cartmill, David Pilbeam, and Glynn Isaac (1986), *American Scientist,* vol.74, pp.410–420.

9 Jane Goodall (1971), *In the Shadow of Man,* Collins, London; (1986), *The Chimpanzees of Gombe,* Harvard University Press, Cambridge.

10 Robert Ardrey (1961), *African Genesis,* Dell, New York; Raymond Dart (1949), *American Journal of Physical Anthropology,* vol.7, pp.1–38; (1953), *International Anthropological and Linguistic Review,* vol.1, pp.201–217. An overview of the hunting hypothesis is presented by Matt Cartmill (1993), *A View of a Death in the Morning,* Harvard University Press, Cambridge.

12 Mythic influences are discussed by Matt Cartmill (1983), *Natural History,* vol.92, pp.65–79; and Misia Landau (1991), *Narratives of Human Evolution,* Yale University Press, New Haven.

14 Daniel Boorstin (1983), *The Discoverers,* Vintage, New York.

II: DOMINION

16 The oldest known bipedal trails are from Laetoli, Tanzania, reported by Mary Leakey and Richard Hay (1979), *Nature,* vol.278, pp.317–323; and M. D. Leakey and J. M. Harris, eds. (1987), *Laetoli,* Clarendon, Oxford.

19–22 The history of Carboniferous and Permian land ecosystems is documented by William DiMichele and Robert Hooke, et al. (1992), "Paleozoic Terrestrial Ecosystems" in *Terrestrial Ecosystems Through Time,* A. K. Behrensmeyer, J. D. Damuth, W. A. DiMichele, R. Potts, H. Sues, and S. L. Wing, eds., University of Chicago Press, Chicago, pp.204–325. Threshold-type change in Carboniferous vegetation, based on research by Bill DiMichele and Tom Phillips, and in early Cenozoic ecosystems, based on research by Scott Wing, is reported by R. Monastersky (1990), *Science News,* vol.138, pp.184–186.

Terrestrial Ecosystems Through Time and extensive references within it also document Mesozoic and Cenozoic ecosystem history. The work of E. C. Olson has been particularly important in uncovering the characteristics of Permian ecosystems on land.

23–24 Agricultural development in northern Khuzistan is documented by Frank Hole, Kent Flannery, and James Neely (1969), "Prehistory and Human Ecology of the Deh Luran Plain," *Museum of Anthropology, University of Michigan, Memoir 1.*

24 The rise of food production in North America is reviewed by Bruce Smith (1989), *Science,* vol.246, pp.1566–1571. For other areas of the world: C. W. Cowan and P. J. Watson (1992), *The Origins of Agriculture,* Smithsonian Press, Washington, D.C. A general discussion of the origin of food production is offered by Richard W. Redding (1988), *Journal of Anthropological Archaeology,* vol.7, pp.56–97; and B. Smith (1994), *The Emergence of Agriculture,* Scientific American Library, New York.

24–25 The history of rice agriculture in Indonesia is recounted by Clifford Geertz (1963), *Agricultural Involution,* University of California Press, Berkeley.

25–28 Carolyn Merchant (1989), *Ecological Revolutions,* University of North Carolina Press, Chapel Hill.

26 The toll of the plague is covered by D. Boorstin (1983), *The Discoverers,* Vintage, New York, p.667.

33–35 The comparison between post-Permian and human-dominated ecosystems was first given by R. Potts, A. K. Behrensmeyer, et al. (1992), "Late Cenozoic Ecosystems" in *Terrestrial Ecosystems Through Time,* A. K. Behrensmeyer, et al., eds., pp.418–541.

36 Mark Plotkin (1988), "The Outlook for New Agricultural and Industrial Products from the Tropics" in *Biodiversity,* E. O. Wilson, ed., National Academy Press, Washington, D.C., p.107; and references cited.

36 Land devoted to food production: G. L. Atjay, P. Ketner, and P. Duvigneaud (1979), "Terrestrial Primary Production and Phytomass" in *The Global Carbon Cycle,* B. Bolin, E. T. Degens, S. Kempe, and P. Ketner, eds., Wiley, New York, pp.129–182.

37–38 Figures on population growth, resource use, wastes, and atmospheric change are from the widely circulated Worldwatch Institute's Report for 1992: Lester R. Brown, et al. (1992), *State of the World 1992,* Norton, New York; the World Resources Institute's volume *World Resources 1992–93,* Oxford University Press, New York; B. L. Turner, et al., eds. (1990), *The Earth as Transformed by Human Action,* Cambridge University Press, New York; C. Mungall and D. J. McLaren, eds., *Planet Under Stress,* Oxford University Press, Toronto; and R. L. Wyman, ed., *Global Climate Change and Life on Earth,* Routledge, Chapman and Hall, New York.

39 Cooling caused by the eruption of Mount Pinatubo is described in P. Minnis, et al. (1993), *Science,* vol.259, pp.1411–1415; and R. A. Kerr (1993), *Science,* vol.259, p.594.

39 Peter Raven (1988), "The Cause and Impact of Deforestation" in *Earth '88: Changing Geographic Perspectives,* National Geographic Society, Washington, D.C., p.225; Raven (1988), "Our Diminishing Tropical Forests" in *Biodiversity,* E. O. Wilson, ed., National Academy Press, Washington, D.C., pp.121.

39 Edward O. Wilson (1992), *The Diversity of Life,* Norton, New York, p.280.

39–40 Population decline and species loss in many groups of animals are reviewed in many books and articles, including the volume edited by Wilson (1988); Turner, et al. (1990); Wilson's book (1992); and Bryan G. Norton (1986), *The Preservation of Species,* Princeton University Press, Princeton.

40 Ape population estimates are reported by Rachel Nowak (1995), *Science,* vol.267, pp.1761–1762, and vol.268, p.25; and from E. Linden (1992), *National Geographic,* March, pp.2–45.

40 Bill McKibben (1989), *The End of Nature,* Anchor, New York.

40 Statistics on habitat loss: John C. Ryan (1992), "Conserving Biological Diversity" in *State of the World 1992,* L. R. Brown, et al., Norton, New York, pp.10–11.

41 Edward O. Wilson (1992), *The Diversity of Life.*

41 The tropical hunter-gatherer model of human evolution is illustrated by R. Lee and I. DeVore, eds. (1968), *Man the Hunter,* Aldine, Chicago; and M. Konner (1982), *The Tangled Wing,* Harper and Row, New York, pp.4–10.

41 Edward O. Wilson (1978), *On Human Nature,* Harvard University Press, Cambridge, Mass., p.196.

42 The quote illustrating the process of evolution comes from Milford H. Wolpoff (1980), *Paleoanthropology,* Knopf, New York, p.22.

42 Richard G. Klein (1989), *The Human Career,* University of Chicago Press, Chicago, p.181.

III: NATURE'S ALTERATION

45 The Lainyamok site has been reported in P. Shipman, R. Potts, and M. Pickford (1983), *Nature,* vol.306, pp.365–368; and R. Potts, P. Shipman, and E. Ingall (1988), *Journal of Human Evolution,* vol.17, pp.597–614.

46–49 The geology of Olorgesailie is presented by Glynn Isaac (1978), "The Olorgesailie Formation" in W. W. Bishop, ed. (1978), *Geological Background to Fossil Man,* Scottish Academic Press, Edinburgh. The stratigraphy discussed in the text comes from personal observations, summarized in my 1994 article in *Journal of Human Evolution,* vol.27, pp.7–24; and from discussions with our project geologist Kay Behrensmeyer. The ages of the ash layers are presented in Alan Deino and R. Potts (1990), *Journal of Geophysical Research,* vol.95, no. B6, pp.8453–8470.

50 The isotope curve is based on K. G. Miller, et al. (1987), *Paleoceanography,* vol.2, pp.1–19; also see M. E. Raymo and W. F. Ruddiman (1992), *Nature,* vol.359, pp.117–122.

50–51 The global climate implications of the oxygen isotope record of ocean foraminifera were first addressed by C. Emiliani (1955), *Journal of Geology,* vol.63, pp.538–578; and reevaluated by N. Shackleton (1967), *Nature,* vol.215, pp.5–17. See Shackleton and N. D. Opdyke (1973), *Quaternary Research,* vol.3, pp.39–55; J. Imbrie and K. P. Imbrie (1979), *Ice Ages,* Harvard University Press, Cambridge, Mass.; and B. J. Skinner and S. C. Porter (1992), *The Dynamic Earth,* Wiley, New York, pp.314–322. The latter volume offers an excellent introduction to the variety of Earth's geological and environmental phenomena.

52 Earth's hydrological cycle and its climatic influences are detailed by M. T. Chahine (1992), *Nature,* vol.359, pp.373–380.

54–59 See review and references in R. Potts, A. K. Behrensmeyer, et al. (1992), "Late Cenozoic Ecosystems" in *Terrestrial Ecosystems Through Time,* Chapter 7. Evidence of a warm, moist vegetation and an equable climate during the early Eocene, and the shift to a cooler, more open, and provincial vegetation worldwide is documented by J. Wolfe (1980), *Palaeogeography, Palaeoclimatology, Palaeoecology,* vol.30, pp.313–323; Wolfe (1985), "Distribution of Major Vegetational Types During the Tertiary" in *The Carbon Cycle and Atmospheric* CO_2, American Geophysical Union Monograph 32, pp.357–375; S. L. Wing (1987), *Annals of the Missouri Botanical Garden,* vol.74, pp.748–784; Wing (1991), *Geology,* vol.18, pp.539–540. The global cooling trend is examined by Eric J. Barron (1985), *Palaeogeography, Palaeoclimatology, Palaeoecology,* vol.50, pp.45–61.

Changes during the late Eocene and early Oligocene are documented in D. R. Prothero and W. A. Berggren, eds. (1992), *Eocene-Oligocene Climatic and Biotic Evolution,* Princeton University Press, Princeton, N.J.

General trends in mammalian evolution, including the evolution of hypsodont teeth and open-habitat locomotion, are presented by Christine M. Janis and John Damuth (1990), "Mammals" in *Evolutionary Trends,* K. J. McNamara, ed., Belhaven, London, pp.301–345.

Evidence of open, grassy vegetation in the Oligocene of Mongolia is noted by J.A.H. Van Couvering (1980), "Community Evolution in East Africa During the Late Cenozoic" in *Fossils in the Making,* A. K. Behrensmeyer and A. Hill, eds., University of Chicago Press, Chicago, pp.272–298.

Saharan history based on deep-ocean dust records is documented by M. Sarnthein (1978), "Neogene Sand Layers off Northwest Africa" in *Initial Reports of the Deep Sea Drilling Project 41,* Y. Lancelot, E. Siebold, et al., U.S. Government Printing Office, Washington, D.C., pp.939–959.

Changes in North American large mammals are reviewed by Richard Stucky (1989), *Current Mammalogy,* vol.2, pp.375–432.

The chronofauna concept was developed by E. C. Olson (1952), *Evolution,* vol.6, pp.181–196; applied to the Miocene of Eurasia by R. L. Bernor (1983), "Geochronology and Zoogeographic Relationships of Miocene Hominoidea" in *New Interpretations of Ape and Human Ancestry,* R. L. Ciochon and R. S. Corruccini, eds., Plenum, New York, pp.21–64; and applied to the Miocene of North America by S. D. Webb (1983), "The Rise and Fall of the Late Miocene Ungulate Fauna in North America" in *Coevolution,* M. H. Nitecki, ed., University of Chicago Press, Chicago, pp.267–306.

60 The importance of parallelisms in mammalian evolution during the late Cenozoic is documented by C. M. Janis and J. Damuth (1990).

60 The expansion of grassland ecosystems in Asia and North America may represent a global change in atmospheric carbon dioxide concentration, according to Thure Cerling, et al. (1993), *Nature,* vol.361, pp.344–345.

60 The term "savanna" refers to a habitat that has a continuous layer of grass interspersed with trees and shrubs. In modern environments, tree cover of more than 20 percent is called a woodland, although grassy environments with 20–50 percent tree cover are often called "savanna-woodland" D. J. Pratt, et al. (1966), *Journal of Applied Ecology,* vol.3, pp.369–382; D. R. Cahoon, et al. (1992), *Nature,* vol.359, p.812).

The "savanna hypothesis" discussed here and in Chapter IV refers to the idea of a directional trend (not necessarily gradual) to a more open habitat, i.e., eventual replacement of forest by grassland, and to the idea that hominids used the resources of, and were otherwise adapted to, the most open habitats within the mosaic of environments at any given time.

62 Environmental diversity: Janis and Damuth (1990), p.307.

63–64 Tectonic uplift during the late Cenozoic and its climatic effects are detailed by William Ruddiman and John Kutzbach (1991), *Scientific American,* March, pp.66–75; Raymo and Ruddiman (1992), *Nature,* vol.359, pp.117–122; Ruddiman and Kutzbach (1989), *Journal of Geophysical Research,* vol.94 (D15), pp.18,409–18,418; Ruddiman, W. L. Prell, and M. E. Raymo

(1989), *Journal of Geophysical Research,* vol.94 (D15), pp.18,379–18,391; Kutzbach, P. J. Guetter, Ruddiman, and Prell (1989), *Journal of Geophysical Research,* vol.94 (D15), pp.18,393–18,407; Raymo, Ruddiman, and P. N. Froelich (1988), *Geology,* vol.16, pp.649–653; and Prell and Kutzbach (1992), *Nature,* vol.360, pp.647–652.

65 Ralph E. Taggart, Aureal T. Cross, and Loretta Satchell, *Proceedings of the Third North American Paleontological Convention,* vol.2, pp.535–540.

65 The history of East African volcanism and rifting is examined in W. W. Bishop, ed. (1978), *Geological Background to Fossil Man,* Scottish Academic Press, Edinburgh; and L. E. Frostick, et al., eds. (1986), *Sedimentation in the African Rifts,* Special Publication of the Geological Society No. 25.

65–66 The Fort Ternan debate can be followed in P. Shipman, et al., (1981), *Journal of Human Evolution,* vol.10, pp.49–72; Shipman (1986), *Journal of Human Evolution,* vol.15, pp.193–204; M. Pickford (1987), *Journal of Human Evolution,* vol.16, 305–309; G. J. Retallack, et al. (1990), *Science,* vol.247, pp.1325–1328; and T. E. Cerling, et al. (1991), *Journal of Human Evolution,* vol.21, pp.295–306 (and references in the latter).

66 The most recent synthesis of the mammalian faunas of the Tugen Hills, Kenya, occurs in Andrew Hill, et al. (1985), *Journal of Human Evolution,* vol.14, pp.759–773. Savanna and open woodland interpretations can be found in Van Couvering (1980), and others cited by Potts, Behrensmeyer, et al. (1992), p.469. The forest paleoflora from Baringo is documented by B. F. Jacobs and C. Kabuye (1987), *Journal of Human Evolution,* vol.16, pp.147–155. The heterogeneity of vegetation during the mid-Miocene of central Kenya is further documented by J. D. Kingston, et al. (1994), *Nature,* vol.264, pp.955–959.

66–67 Climatic and vegetational changes in Africa during the mid-and late Miocene are addressed by D. I. Axelrod and P. H. Raven (1978), "Late Cretaceous and Tertiary Vegetation History of Africa" in *Biogeography and Ecology of Southern Africa,* M.J.A. Werger, ed., D. W. Junk, The Hague, pp.77–130; K. W. Butzer and H.B.S. Cooke (1982), "The Palaeo-ecology of the African Continent" in *The Cambridge History of Africa,* vol.1, J. D. Clark, ed., Cambridge University Press, Cambridge, Eng.; L. Laporte and A. Zihlman (1983), *South African Journal of Science,* vol.79, pp.96–110; R. Bonnefille (1985), *South African Journal of Science,* vol.81, pp.267–270; E. M. Van Zinderen Bakker and J. H. Mercer (1986), *Palaeogeography, Palaeoclimatology, and Palaeoecology,* vol.56, pp.217–235; and Potts, Behrensmeyer, et al. (1992), pp.464–471.

67 The history of Antarctic glaciation is reviewed by G. Robin (1988), *Palaeogeography, Palaeoclimatology, Palaeoecology,* vol.67, pp.31–50; P.-N. Webb (1990), *Antarctic Science,* vol.2, pp.3–21; and P. J. Barrett, et al. (1987), *Geology,* vol.15, pp.634–637.

68 Antarctic climatic instability is discussed by P.-N. Webb and D. M. Har-

wood (1991), *Quaternary Science Reviews,* vol.10, pp.215–223; P. J. Barrett, et al. (1992), *Nature,* vol.359, pp.816–818 (and references therein); reviewed by D. Sugden (1992), *Nature,* vol.359, pp.775–776; and disputed by D. R. Marchant, et al. (1993), *Science,* vol.260, pp.667–670.

69–71 Orbital cycles, their effects on environmental variability, and the tempo of fluctuation are examined by J. D. Hays, J. Imbrie, and N. J. Shackleton (1976), *Science,* vol.194, pp.1121–1132; Imbrie and Imbrie (1979); and Skinner and Porter (1992). See note for pp.50–51.

H.-S. Liu (1992) offers an analysis suggesting that the 100,000-year climatic cycle may be caused by changes in Earth's obliquity (tilt) rather than eccentricity (orbital shape change); *Nature,* vol.358: 397–399.

71–72 The skeletal anatomy of *Proconsul* is scrutinized by Carol Ward, Alan Walker, and colleagues (1993), *American Journal of Physical Anthropology,* vol.90, pp.77–111; Ward (1993), *American Journal of Physical Anthropology,* vol.92, pp.291–328; also see Michael D. Rose (1983), "Miocene Hominoid Postcranial Morphology: Monkey-like, Ape-like, Neither, or Both?" in *New Interpretations of Ape and Human Ancestry,* R. L. Ciochon and R. S. Corruccini, eds., Plenum, New York, pp.405–417.

72 Environments of early and mid-Miocene apes are discussed by P. Andrews and J. H. Van Couvering (1975), "Palaeoenvironments in the East African Miocene" in *Approaches to Primate Paleobiology,* F. S. Szalay, ed., Karger, Basel, pp.62–103; Andrews (1992), *Journal of Human Evolution,* vol.22, pp.423–438; and Andrews (1992), *Nature,* vol.360, pp.641–646.

72–75 Miocene ape diversity and relationships are discussed by David Begun (1992), *Science,* vol.257, pp.1929–1933; and Peter Andrews (1992), *Nature,* vol.360, pp.641–646.

73 The directional turn (cooling) in the oxygen isotope record at 15 million years ago is an average that is well represented in the diagram on p.50. The more detailed curve, which shows repeated oscillation from 22 to 12 million years ago is from F. Woodruff, S. M. Savin, and R. G. Douglas (1981), *Science,* vol.212, p.666.

74 Ape dietary change during the Miocene is reviewed by P. Andrews and L. Martin (1991), *Philosophical Transactions of the Royal Society, London,* B, vol.334, pp.199–209.

75–76 The late Miocene rise of chronofaunas and their disruption is addressed by S. D. Webb (1989), *Annual Reviews of Earth and Planetary Sciences,* vol.16, pp.413–438; Webb (1989), "The Fourth Dimension in North American Terrestrial Mammal Communities" in *Patterns in the Structure of Mammalian Communities,* D. W. Morris, et al., eds., Special Publication No. 28, Museum, Texas Tech University, pp.181–203; Bernor (1983), pp.50–54; and V. J. Maglio (1978), "Patterns in Faunal Evolution" in *Evolution of African Mammals,* Maglio and H.B.S. Cooke, eds., Harvard University Press, Cambridge, pp.603–619.

76 Oxygen isotope data for climatic variations during the late Miocene are reported by R. Stein and M. Sarnthein (1984), *Palaeoecology of Africa,* vol.16, pp.9–36.

76–77 The Mediterranean salinity crisis, known as the "Messinian event," was first reported as if it had been a single episode 5.5 million years ago, leading to savanna expansion in Africa; see K. J. Hsü, et al. (1977), *Nature,* vol.267, pp.399–403. But according to K. J. Hsü, et al. (1978), *Leg 42, Deep Sea Drilling Project Initial Report,* vol.42, pp.1053–1078, the crisis involved multiple large-scale flooding and drying events. These fluctuations occurred between 6.4 and 4.6 million years ago, according to Stein and Sarnthein (1984). See reviews by Laporte and Zihlman (1983), p.105; and Richard Cowan (1995), *History of Life,* Blackwell, Boston, pp.383–385. Recent work by W. A. Berggren suggests that the Messinian oscillations began as early as 7 million years ago (Raymond Bernor, personal communication).

IV: EXPERIMENTS IN BEING HUMAN

79 The possible earliest known hominids from the late Miocene and earliest Pliocene of Africa are discussed by A. Hill and S. Ward (1988), *Yearbook of Physical Anthropology,* vol.31, pp.49–83; and Hill (1985), *Nature,* vol.315, pp.222–224. Also see recent articles in *National Geographic,* November 1995.

79 T. D. White, et al. (1994), *Nature,* vol.371, pp.306–312. White, et al. (1995) have recently proposed that the early hominids of Aramis be assigned to a new hominid genus, *Ardipithecus;* see *Nature,* vol.375, p.88.

79 M. G. Leakey, et al. (1995), *Nature,* vol.376, pp.565–571.

80 The Laetoli footprints, fossils, and their context: Leakey and Hay (1979); Leakey and Harris, eds. (1987).

80 The Hadar site and skeleton AL-288 ("Lucy") are discussed by D. C. Johanson, M. Taieb, and Y. Coppens (1982), *American Journal of Physical Anthropology,* vol.57, p.373–402, and other articles in that same journal issue; Kimbel, et al. (1994), *Nature,* vol.368, pp.449–451; and, for the more general reader, D. C. Johanson and M. Edey (1981), *Lucy: The Beginnings of Humankind,* Simon and Schuster, New York.

80–81 C. O. Lovejoy (1978), "A Biomechanical Review of the Locomotor Diversity of Early Hominids" in *Early Hominids of Africa,* C. J. Jolly, ed., St. Martin's, New York, pp.403–429; Lovejoy, et al. (1979), *American Journal of Physical Anthropology,* vol.57, pp.679–700; Johanson and Edey (1981).

81 J. T. Stern and R. L. Susman (1983), *American Journal of Physical Anthropology,* vol.60, pp.279–317; Susman, et al. (1984), *Folia Primatologica,* vol.43, pp.113–156; W. L. Jungers (1982), *Nature,* vol.297, pp.676–678. Also see comments by M. H. Wolpoff (1983), *Nature,* vol.304, pp.59–61; Jungers

and Stern (1983), *Journal of Human Evolution,* vol.12, pp.673–684; and D. L. Gebo (1992), *American Journal of Physical Anthropology,* vol.89, pp.29–58.

81 B. Latimer and C. O. Lovejoy (1989, 1990), *American Journal of Physical Anthropology,* vol.78, pp.369–386; and vol.82, pp.125–133; Latimer, et al. (1987), vol.74, pp.155–175. Debate about the Laetoli footprints: M. H. Day and E. H. Wickens (1980), *Nature,* vol.286, pp.385–387; T. D. White and G. Suwa (1987), *American Journal of Physical Anthropology,* vol.72, pp.485–514; compare with the articles noted above by Stern, Susman, and Jungers.

81 R. J. Clarke and P. V. Tobias (1995), *Science,* vol.269, pp.521–524.

82–84 The savanna hypothesis is reiterated in various forms in virtually all texts on human evolution; the importance accorded to the savanna adaptation is illustrated by Laporte and Zihlman (1983), pp.96–99, 105–107.

83 S. L. Washburn (1960), *Scientific American,* September; K. P. Oakley (1961), *Man the Toolmaker,* University of Chicago, Chicago.

84 Foraminifera-isotope summary is based on Stein and Sarnthein (1984).

86 Aramis paleoenvironment is based on G. WoldeGabriel, et al. (1994), *Nature,* vol.371, pp.330–333.

86–87 Laetoli paleoenvironment is based on articles in Leakey and Harris (1987), and P. Andrews (1989), *Journal of Human Evolution,* vol.18, pp.173–181.

87 The fluctuation of habitats at Hadar is based on Johanson, et al. (1982), pp.379–380; T. Gray (1980), "Environmental Reconstruction of the Hadar Formation," Ph.D. thesis, Case Western Reserve University, Cleveland; and R. Bonnefille, et al. (1987), *Palaeogeography, Palaeoclimatology, Palaeoecology,* vol.60, pp.249–281.

87–88 Alternative interpretations of Makapansgat environments are given by King (1951), *Transactions of the Royal Society of South Africa,* vol.33, pp.121–151; C. K. Brain (1958), *Transvaal Museum Memoir* No. 11; K. W. Butzer (1971), *American Anthropologist,* vol.73, pp.1197–1201; B. R. Turner (1980), *Palaeontologia Africana,* vol.23, pp.51–58; T. C. Partridge (1980), *Palaeontologia Africana,* vol.23, p.45; G. deGraaf (1960), *Palaeontologia Africana,* vol.7, pp.59–118; J. W. Kitching (1980), *Palaeontologia Africana,* vol.23, pp.63–68; R. F. Ewer (1958), *Proceedings of the Zoological Society London,* vol.130, pp.329–372; and E. S. Vrba (1988), "Late Pliocene Climatic Events and Hominid Evolution" in *Evolutionary History of the "Robust" Australopithecines,* F. Grine, ed., Aldine de Gruyter, New York, pp.405–426.

88 A. Cadman and R. J. Rayner (1989), *Journal of Human Evolution,* vol.18, pp.107–113. The adaptation-to-forest viewpoint is adopted by Rayner, et al. (1993), *Journal of Human Evolution,* vol.24, pp.219–231.

88 An early australopith jaw, 3 to 3.5 million years old, has recently been announced from Chad, central Africa. Fossil animals from the site indicate forested, wooded savanna, and grassy habitats. M. Brunet, et al. (1995), *Nature,* vol.378, pp.273–275.

88– R. Bonnefille (1984), "Cenozoic Vegetation and Environments of Early
89 Hominids in East Africa" in *The Evolution of the East Asian Environment,* vol.2, R. O. Whyte, ed., Centre of Asian Studies, Hong Kong, pp.597–598; P. G. Williamson (1985), *Nature,* vol.315, pp.487–489.

91 A. Hill (1987), *Journal of Human Evolution,* vol.16, p.593.

92– The prevalence of parallelisms during the late Cenozoic is documented by
95 Janis and Damuth (1990), and by Potts, Behrensmeyer, et al. (1992).

93 S. J. Gould (1985), *The Flamingo's Smile,* Norton, New York, p.210; Gould (1989), *Wonderful Life,* Norton, New York.

94 The land-bridge exchange between South and North America is a dramatic case documented by S. D. Webb (1976), *Paleobiology,* vol.2, pp.220–234; and L. G. Marshall (1988), *American Scientist,* vol.96, pp.380–388.

97 R. Wrangham (1980), *Journal of Human Evolution,* vol.9, pp.329–331.

98 P. Wheeler (1991), *Journal of Human Evolution,* vol.21, pp.107–136.

99 M. D. Leakey (1971), *Olduvai Gorge,* vol.3, Cambridge University Press, Cambridge, Eng., p.278.

101– The fauna of the period from 3 to 1.5 million has been documented by J.
102 M. Harris, ed. (1983, 1991), *Koobi Fora Research Project,* vols. 2 and 3, Clarendon, Oxford, Eng.; V. J. Maglio and H.B.S. Cooke, eds. (1978), *Evolution of African Mammals,* Harvard University Press, Cambridge, Mass.; Y. Coppens, et al., eds. (1976), *Earliest Man and Environments in the Lake Rudolf Basin,* University of Chicago, Chicago; C. K. Brain (1981), *The Hunters or the Hunted?,* University of Chicago, Chicago.

102– Hominid species diversity is documented by B. A. Wood (1991), *Koobi Fora*
106 *Research Project,* vol.4, Clarendon, Oxford, Eng.; Wood (1992), *Nature,* vol.355, pp.783–790; Wood (1994), *Nature,* vol.371, p.280; D. C. Johanson, et al. (1987), *Nature,* vol.327, pp.205–209; A. Walker, et al. (1986), *Nature,* vol.322, pp.517–522; G. P. Rightmire (1993), *American Journal of Physical Anthropology,* vol.90, pp.1–34; and articles by R. J. Clarke and others in *Evolutionary History of the "Robust" Australopithecines,* F. Grine, ed. (1988), Aldine de Gruyter, New York.

105 Body sizes of early hominids are estimated by Henry McHenry (1992), *American Journal of Physical Anthropology,* vol.87, pp.407–431.

105 The chewing apparatus and microscopic tooth wear of the robust australopiths indicates the influence of hard or tough plant foods: see articles by W. Hylander, and R. Kay and F. Grine in *Evolutionary History of the "Robust" Australopithecines,* F. Grine, ed. Recent research by Andrew Sillen indicates that there was also an animal component in the diet of the robust australopiths of Swartkrans, South Africa: Sillen (1992), *Journal of Human Evolution,* vol.23, pp.495–516; and Sillen, et al. (1995), *Journal of Human Evolution,* vol.28, pp.277–286.

107– The oldest archeological sites, stone tools, and their manufacture are de-
109 scribed by M. D. Leakey (1971); G. L. Isaac (1984), *Advances in World*

Archaeology, vol.3, pp.1–87; N. Toth and K. Schick (1986), *Advances in Archaeological Method and Theory,* vol.9, pp.1–96; R. Potts (1991), *Journal of Anthropological Research,* vol.47, pp.153–176; and Kathy Schick and Nicholas Toth (1993), *Making Silent Stones Speak,* Simon and Schuster, New York.

107–109 Cut marks on bones from Oldowan sites were first reported by R. Potts and P. Shipman (1981), *Nature,* vol.291, pp.577–580; and H. T. Bunn (1981), *Nature,* vol.291, pp.574–577.

107–108 Hominid transport of resources to Oldowan sites and comparisons with chimp nut-crackers are explored in my book *Early Hominid Activities at Olduvai,* Aldine de Gruyter, New York (1988). Movements by nut-cracking chimps are described by C. and H. Boesch (1984), *Primates,* vol.25, pp. 160–170.

110 L. H. Keeley and N. Toth (1981), *Nature,* vol.293, pp.464–465.

111 R. J. Blumenschine (1986), *Early Hominid Scavenging Opportunities,* British Archaeological Reports International Series 283; Blumenschine and J. A. Cavallo (1992), *Scientific American,* April, pp.90–96.

112 R. Potts (1984), *American Scientist,* vol.72, pp.338–347; Potts (1988); Potts (1994), *Journal of Human Evolution,* vol.27, pp.7–24.

113 The 2.4- to 2.5-million-year-old event is documented by N. Shackleton, et al. (1984), *Nature,* vol.307, pp.620–623.

113 Loess deposition in central China is reviewed by George Kukla and Zhisheng An (1989), *Palaeogeography, Palaeoclimatology, Palaeoecology,* vol.72, pp. 203–225.

113 Henry Hooghiemstra's study is reported by E. S. Vrba (1988), p.408.

114–115 The turnover-pulse idea is advanced by Vrba (1985), "Ecological and Adaptive Changes Associated with Early Hominid Evolution" in *Ancestors: The Hard Evidence,* E. Delson, ed., Liss, New York, pp.63–71; Vrba (1985), *South African Journal of Science,* vol.81, pp.229–236; Vrba (1988); Vrba, G. H. Denton, and M. L. Prentice (1989), *Ossa,* vol.14, pp.127–156; and Vrba (1992), *Journal of Mammalogy,* vol.73, pp.1–28.

115 Directional environmental change, or climatic forcing—the importance of the spread of savannas due to global cooling at 2.4 million years ago—is also key to the hypothesis of human evolution advanced by Steven M. Stanley (1992), *Paleobiology,* vol.18, pp.237–257; (1995), "Climatic Forcing and the Origin of the Human Genus" in *Effects of Past Global Change on Life,* National Academy Press, Washington, D.C., pp.233–243.

116 Shackleton, et al. (1984); diagram represents a portion of the curve for Atlantic Ocean Site 552A. For reference on Pacific core, see Figure 4 in the Shackleton paper.

117 Stein and Sarnthein (1984); curves for Atlantic Ocean Sites 397 and 141 (p.15) illustrate the widened pattern of oxygen isotopic variation between 3.5 and 1.9 million years ago.

117 W. F. Ruddiman, et al. (1989), *Proceedings of the Ocean Drilling Program, Scientific Results,* vol.108, pp.474, 478.

117 H. Hooghiemstra (1989), *Palaeogeography, Palaeoclimatology, Palaeoecology,* vol.72, pp.11–26; Hooghiemstra and E. Ran (1994), *Quaternary International,* vol.21, pp.63–80.

117–118 Craig Feibel, Frank Brown, and John M. Harris (1991), "Neogene Paleoenvironments of the Turkana Basin" in *Koobi Fora Research Project,* vol.3, J. M. Harris, ed., Clarendon, Oxford, Eng., pp.321–370.

118 Pollen change: R. Bonnefille (1985), *South African Journal of Science,* vol.81, p.269. Change in fossil mammals: J. M. Harris, ed. (1983, 1991), *Koobi Fora Research Project,* vols.2 and 3. Isotopic data from paleosols also weaken the idea of a single, permanent replacement of relatively closed vegetation by open savanna. In the Turkana sequence, the main shift to C4 vegetation (grasses) is not detected until after 1.8 million years (T. Cerling, et al. [1988], *Palaeogeography, Palaeoclimatology, Palaeoecology,* vol.63, pp.335–356). In the Tugen Hills sequence, the shift away from C3 vegetation (forest and woodland) is not recorded until after 2 million years ago (the upper Chemeron Formation; see J. D. Kingston, et al. [1994], *Science,* vol.264, p.957).

118 The Black Skull: A. Walker, et al. (1986); papers in Grine, ed. (1988).

118–119 Oldest archeological occurrences in the Turkana basin: M. Kibunjia (1994), *Journal of Human Evolution,* vol.27, pp.159–171. Malawi: Z. M. Kaufulu and N. Stern (1987), *Journal of Human Evolution,* vol.16. pp.729–740. Zaire: J.W.K. Harris, et al. (1987), *Journal of Human Evolution,* vol.16, pp.701–728. Reviewed by J.W.K. Harris (1983), *African Archaeological Review,* vol.1, pp.3–31.

119 Baringo fossil: A. Hill, et al. (1992), *Nature,* vol.355, pp.719–722.

119–120 The "broad pulse" idea is reported by W. K. Stevens, "Dust in Sea Mud May Link Human Evolution to Climate," *The New York Times,* December 14, 1993, p.C1; and the related work of Peter deMenocal is reported by R. A. Kerr (1994), *Science,* vol.263, pp.173–174.

119–120 R. Foley (1987), *Another Unique Species,* Longman, Essex, Eng., pp.247–248.

120 M. L. Prentice and G. H. Denton (1988), "The Deep-Sea Oxygen Isotope Record, the Global Ice Sheet System and Hominid Evolution" in *Evolutionary History of the "Robust" Australopithecines,* F. Grine, ed., Aldine de Gruyter, New York, pp.383–403; quote on p.398.

120 A "broad pulse" model, with climatic drying starting 2.8 million years ago, is also presented by Peter deMenocal (1995), *Science,* vol.270, pp.53–59. Despite this paper's emphasis on greater aridity in East Africa, deMenocal's data actually show a change in the frequency and a heightened degree of fluctuation.

122 Ocean oxygen isotope fluctuations between 2.2 and 2 million years ago can

be observed in the curves published by Shackleton, et al. (1984), p.622; and by Stein and Sarnthein (1984), p.15.

122 Environmental fluctuations at Olduvai are documented by R. L. Hay (1976), *Geology of the Olduvai Gorge,* University of California, Berkeley; R. Bonnefille (1984), *National Geographic Research Reports,* vol.17, pp.227–243; Potts (1988); and R. C. Walter, et al. (1991), *Nature,* vol.354, pp.145–149.

125 A. Walker and R. Leakey, eds. (1993), *The Nariokotome Homo erectus Skeleton,* Harvard University Press, Cambridge, Mass.

126 L. C. Aiello and P. Wheeler (1995), *Current Anthropology,* vol.36, pp. 199–221.

128 Dmanisi: L. Gabunia and A. Vekua (1995), *Nature* vol.373, pp.509–512. Java: C. C. Swisher, et al. (1994), *Science,* vol.263, pp.1118–1121; J. de Vos and P. Sondaar (1994), *Science,* vol.266, pp.1726–1727. 'Ubeidiya: E. Tchernov (1988), *Paléorient,* vol.14, pp.63–65. China: Kathy Schick and Dong Zhuan (1993), *Evolutionary Anthropology,* vol.2, pp.22–23; Huang Wanpo, et al. (1995), *Nature,* vol.378, pp.275–278. Ethiopia: J. D. Clark and H. Kurashina (1979), *Nature,* vol.282, pp.33–39. Europe: W. Roebroeks (1994), *Current Anthropology,* vol.35, pp.301–305; E. Carbonell, et al. (1995), *Science,* vol.269, pp.826–830.

129 Alan Turner (1984), "Hominids and Fellow-Travellers" in *Hominid Evolution and Community Ecology,* R. Foley, ed., Academic Press, London, pp.193–217.

130 C. B. Ruff (1991), *Journal of Human Evolution,* vol.21, pp.81–105; Ruff and A. Walker (1993), "Body Size and Body Shape" in *The Nariokotome Homo erectus Skeleton,* A. Walker and R. Leakey, eds. (1993), pp.234–265.

130 Skin: W. Montagna (1985), *Journal of Human Evolution,* vol.14, pp.3–22.

130 Nasal morphology in early hominids: John Gurche, personal communication; R. G. Franciscus and E. Trinkaus (1988), *American Journal of Physical Anthropology,* vol.75, pp.517–527.

131 The development of dry grassland-dominated habitats in Africa is summarized by T. E. Cerling (1992), *Palaeogeography, Palaeoclimatology, Palaeoecology,* vol.97, pp.241–247.

132 The parallels between the robust australopiths and large-bodied, herbivorous mammals are detailed by A. Turner and B. Wood (1993), *Journal of Human Evolution,* vol.24, pp.301–318.

134–135 The degree to which the robust australopiths were committed to "hypergrowth" of tooth enamel and a massive facial architecture related to chewing is documented in articles by F. Grine and L. Martin, Y. Rak, and W. Hylander in *Evolutionary History of the "Robust" Australopithecines,* F. Grine, ed. (1988).

134–135 Extinction of the robust australopiths and the idea of competition with contemporaneous monkeys are discussed by R. Foley (1987), pp.227–255.

V: SURVIVAL OF THE GENERALIST

139 Microscopic studies of handaxe usage have been carried out by Lawrence Keeley (1980), *Experimental Determination of Stone Tool Use,* University of Chicago Press, Chicago.

141 Hypotheses about the rise of innovation and stylistic boundaries during the late Pleistocene: G. Isaac (1972), "Chronology and the Tempo of Cultural Change During the Pleistocene" in *Calibration of Hominid Evolution,* W. W. Bishop and J. A. Miller, eds., Scottish Academic Press, Edinburgh, pp. 399–401.

143 Pleistocene lake-level curve for Olorgesailie is from Potts (1994), p.22.

147 The evolutionary history of *Theropithecus* is documented in the volume edited by Nina Jablonski (1993), *Theropithecus: The Rise and Fall of a Primate Genus,* Cambridge University Press, New York.

148–151 The evolutionary histories of African elephants, pigs, hippos, giraffes, and zebras are documented in Maglio and Cooke, eds. (1978); and Harris, ed. (1983, 1991). The relatively versatile behavior of the common hippopotamus is noted by Harris (1991), pp.57–59; and Hans Klingel (1995), *Natural History,* May, pp.46–57. The versatility of the living giraffe: Harris (1991), p.107. The dietary and social flexibility of the Grévy's zebra: J. Ginsberg (1989) in *The Biology of Large African Mammals in Their Environment,* P. A. Jewell and G.M.O. Maloiy, eds., Clarendon, London.

152–153 Information on Lainyamok and irregular changes in biotic communities: R. Potts and A. Deino (1995), *Quaternary Research,* vol.43, pp.106–113; Potts, et al. (1988).

155 Isotope diagram is from T. J. Crowley and G. R. North (1991), *Paleoclimatology,* Oxford University Press, New York, p.113.

157 N. J. Shackleton (1987), *Quaternary Science Reviews,* vol.6, pp.183–90.

157 Oxygen isotope diagram is based on Crowley and North (1991), p.112.

158 The idea of equilibrium states and the record of sea-level change: G. de Q. Robin (1988), *Palaeogeography, Palaeoclimatology, Palaeoecology,* vol.67, pp. 31–50.

159 Devil's Hole study: T. B. Coplen, I. J. Winograd, et al. (1994), *Science,* vol.263, pp.361–365.

160 M. R. Rampino and S. Self (1992), *Nature,* vol.359, pp.50–52.

160–161 Landscape and biotic changes during glacial and interglacial times: A. J. Sutcliffe (1985), *On the Tracks of Ice Age Mammals,* Harvard University Press, Cambridge, Mass.; E. C. Pielou (1991), *After the Ice Age,* University of Chicago Press, Chicago.

161 Loess-soil curve based on George Kukla (1987), *Quaternary Science Reviews,* vol.6, p.211; and see Kukla and Zhisheng An (1989).

161–162 Grande Pile: G. M. Woillard (1978), *Quaternary Research,* vol.9, pp.1–21; Woillard and W. G. Mook (1982), *Science,* vol.215, pp.159–161.

162 Tenaghi Phillipon: The work of T. A. Wijmstra and his colleagues is summarized by G. Kukla (1989), *Palaeogeography, Palaeoclimatology, Palaeoecology,* vol.72, p.6. Other long pollen sequences in Europe that document large fluctuations in trees, grasses, and ferns during the Pleistocene include: J. Turon (1984), *Nature,* vol.309, pp.673–676; B. W. Sparks and R. G. West (1972), *The Ice Age in Britain,* Methuen, London.

162–163 Long records of dust and terrestrial clays in the eastern Atlantic Ocean: W. Ruddiman, M. Sarnthein, et al. (1989), *Proceedings of the Ocean Drilling Program, Scientific Results,* vol.108, pp.463–484. Windblown lake diatom record: E. M. Pokras and A. C. Mix (1985), *Quaternary Research* vol.24, pp.137–149; (1987), *Nature* vol.326, pp.486–487.

163–164 Last glacial-interglacial vegetation and lake-level change in Africa: E. A. Street and A. T. Grove (1976), *Nature,* vol.261, pp.385–390; (1979), *Quaternary Research,* vol.12, pp.83–118; A. C. Hamilton (1982), *Environmental History of East Africa,* Academic Press, London; J. E. Kutzbach and F. A. Street-Perrott (1985), *Nature,* vol.317, pp.130–134; and M. Sarnthein (1978), *Nature,* vol.272, pp.43–46.

Maps are based on Neil Roberts (1984), "Pleistocene Environments in Time and Space" in *Hominid Evolution and Community Ecology,* R. Foley, ed., p.40.

164–165 Mediterranean sapropels: M. Rossignol-Strick (1983), *Nature,* vol.304, pp.46–49.

165–166 Africa: J. Kingdon (1989), *Island Africa,* Princeton University Press, Princeton. South America: J. Platt Bradbury, et al. (1981), *Science,* vol.214, pp.1299–1305; K-b. Liu and P. A. Colinvaux (1985), *Nature,* vol.318, pp.556–557; and Colinvaux (1989), *Nature,* vol.340, pp.188–189; J. R. Flenley (1979), *The Equatorial Rain Forest: A Geological History,* Butterworth, London; J. Tricart cited in Potts, Behrensmeyer, et al. (1992), p.501 ("pioneer community").

167 The rate of vegetation change during the Pleistocene: A. Traverse (1982), *Alcheringa,* vol.6, pp.197–209; previous notes on Grande Pile and Tenaghi Phillipon pollen records; G. M. MacDonald, et al. (1993), *Nature,* vol.361, pp. 243–246; J. A. Dorale, et al. (1992), *Science,* vol.258, pp.1626–1630; and A. J. Sutcliffe (1985), p.69: Sutcliffe estimates that the change from temperate forest to glacial taiga occurred in about 150 years in the Grande Pile, France.

167–168 Antarctic ice core record: J. Jouzel (1993), *Nature,* vol.364, pp.407–412. Greenland ice core records: S. J. Johnsen, et al. (1992), *Nature,* vol.359, pp.311–313; Grip Members (1993), *Nature,* vol.364, pp.203–207; W. Dansgaard, et al., *Nature,* vol.364, pp.218–220; and the idea of the "flickering switch" comes from K. C. Taylor, et al. (1993), *Nature,* vol.361, pp.432–436. The controversy is highlighted in L. D. Keigwin, et al. (1994), *Nature,* vol.371, pp.323–326; J. F. McManus, et al. (1994), *Nature,* vol.371, pp.326–329; C. D. Charles, et al. (1994), *Science,* vol.263, p.511.

At least one record, obtained from the northwest coast of North America, where fossil pollen and ocean isotopes can be studied together, suggests that considerable fluctuation in vegetation occurred from 128,000 to 118,000 years ago; see L. E. Heusser and N. J. Shackleton (1979), *Science,* vol.204, pp.837–839.

168–172 The review of animal behavior and ecology is based on J. R. Krebs and N. B. Davies, eds. (1984), *Behavioral Ecology,* Sinauer, Sunderland, Mass.; Krebs and Davies (1981), *An Introduction to Behavioural Ecology,* Blackwell, Oxford, Eng.; A. C. Kamil and T. D. Sargent, eds. (1981), *Foraging Behavior,* Garland, New York; D. I. Rubenstein and R. W. Wrangham, eds. (1986), *Ecological Aspects of Social Evolution,* Princeton University Press, Princeton, N.J.; J. Alcock (1989), *Animal Behavior,* Sinauer, Sunderland, Mass.; B. B. Smuts, et al., eds. (1987), *Primate Societies,* University of Chicago Press, Chicago.

174 "Cultural man" quote is from Lee and DeVore, eds. (1968), p.3.

176 Among Glynn Isaac's many influential papers is Isaac (1978), *Scientific American,* April. For a review of the home-base issue, see Potts (1984, 1988).

176 S. L. Washburn and I. DeVore (1961), "Social Behavior of Baboons and Early Man" in *Social Life of Early Man,* Washburn, ed., Viking Fund, New York, vol.31, pp.91–105.

179 A brief, careful summary of the evidence for human control of fire and shelter building is provided by R. Klein (1989).

183–184 Imo's story: S. Kawamura (1959), *Primates,* vol.2, pp.43–60; M. Kawai (1965), *Primates,* vol.6, pp.1–30; H. Kummer (1971), *Primate Societies,* Aldine-Atherton, Chicago; T. Nishida (1987) "Local Traditions and Cultural Transmission" in Smuts, et al., eds. (1987), pp.462–474.

Group differences in grooming, termiting, and other evidence of behavioral traditions in nonhuman primates: R. W. Wrangham, et al., eds. (1994), *Chimpanzee Culture,* Harvard University Press, Cambridge, Mass.; W. C. McGrew (1992), *Chimpanzee Material Culture,* Cambridge University Press, Cambridge, Eng.; M. P. Ghiglieri (1988), *East of the Mountains of the Moon,* Free Press, New York; J. Goodall (1986).

185 Illustration based on photograph by W. C. McGrew.

186 L.S.B. Leakey, P. V. Tobias, J. R. Napier (1964), *Nature,* vol.202, pp.308–312. See Tobias (1991), *Olduvai Gorge,* vol.4, Cambridge University Press, Cambridge, Eng.

187–188 Critique of the handyman hypothesis: Potts (1991); Potts (1993), "Archeological Interpretations of Early Hominid Behavior and Ecology" in *The Origin and Evolution of Humans,* T. D. Rasmussen, ed., Jones and Bartlett, Boston, pp.49–74; N. Toth (1985), *Journal of Archaeological Science,* vol.12, pp.101–120; Schick and Toth (1993); T. Wynn (1981), *Journal of Human Evolution,* vol.10, pp.529–541.

200–203 An overview of late Pleistocene technology, behavior, and art is presented by R. Klein (1989); Klein (1992) *Evolutionary Anthropology,* vol.1, pp.5–14;

P. Mellars and C. Stringer, eds. (1989), *The Human Revolution,* Edinburgh University Press, Edinburgh; L. A. Schepartz (1993), *Yearbook of Physical Anthropology,* vol.36, pp.91–126; H. Knecht, A. Pike-Tay, R. White, eds. (1993), *Before Lascaux,* CRC Press, Boca Raton, Fla.; White (1993), *Natural History,* May, pp.61–67; J. Pfeiffer (1990), *Mosaic,* vol.21, pp.14–23; L. G. Straus (1991) *Journal of Anthropological Research,* vol.47, pp.259–278; O. Soffer and C. Gamble, eds. (1990), *The World at 18,000 B.P.,* vol.1, Unwin Hyman, London.

Recent discovery in the Semliki Valley, Zaire, of shaped bone points used for fishing, about 89,000 years old, suggests the presence of a modern human subsistence strategy by this date; see J. E. Yellen, et al. (1995), *Science,* vol.268, pp.553–556; A. S. Brooks, et al. (1995), *Science,* vol.268, pp.548–553.

205 "Mentality disintegrates": I. B. Black (1992), *Natural History,* February, p.71.

206 Brain growth and energy costs: R. E. Passingham (1985), *Brain, Behavior and Evolution,* vol.26, pp.167–175; S. M. Stanley (1992), *Paleobiology,* vol.18, pp.237–257; Aiello and Wheeler (1995).

207 N. K. Humphrey (1976), "The Social Function of Intellect" in *Growing Points in Ethology,* P.P.G. Bateson and R. A. Hinde, eds., Cambridge University Press, Cambridge, Eng., pp.303–317; A. Jolly (1966), *Science,* vol.153, pp.501–506.

208 R. Byrne and A. Whiten, eds. (1988), *Machiavellian Intelligence,* Oxford University Press, Oxford, Eng.

212 A. Azzaroli (1983), *Palaeogeography, Palaeoclimatology, Palaeoecology,* vol.44, pp.117–139.

213 The quote about "being released from the shackles of environment" comes from C. Stringer and C. Gamble (1993), *In Search of the Neanderthals,* Thames and Hudson, New York, p.177.

213 Pleistocene uplift of the Tibetan plateau: Han Tonglin (1991), "Quaternary Geology of the Qinghai-Tibet (Xizang) Plateau" in *The Quaternary of China,* Zhang Zonghu, ed., China Ocean Press, Beijing, pp.405–440.

214 European data on large mammal extinctions come from Azzaroli (1983); the Java data are from the work of J. de Vos and P. Sondaar, and M. Aimi and F. Aziz (1985) in *Quaternary Geology of the Hominid Fossil Bearing Formations in Java,* N. Watanabe and D. Kadar, eds., Geological Research and Development Centre, Bandung, pp.155–198; the Chinese data are from Xue Xiangxi and Zhang Yunxiang (1991), "Quaternary Mammalian Fossils and Fossil Human Beings" in Zhang Zonghu, ed., pp.307–374.

215 Giant gelada monkey site at Olorgesailie: Isaac (1977); P. Shipman, W. Bosler, and K. L. Davis (1981), *Current Anthropology,* vol.22, pp.257–268; C. Koch (1986), Ph.D. dissertation, University of Toronto; R. Potts, personal observations.

217 Li Tianyuan and D. Etler (1992), *Nature,* vol.357, pp.404–407.

217–218 Three excellent books on Neanderthals for the general audience have come out in recent years: E. Trinkaus and P. Shipman (1993), *The Neandertals,* Knopf, New York; Stringer and Gamble (1993); and J. Shreeve (1995), *The Neandertal Enigma,* Morrow, New York. The long-standing idea of Neanderthal cold adaptations is confirmed in these books and by C. Ruff (1993), *Evolutionary Anthropology,* vol.2, pp.53–60.

219 The Neanderthal pelvis from Kabara: Y. Rak (1990), *American Journal of Physical Anthropology,* vol.81, pp.323–332.

VI: A NEW VIEW OF NATURE

225 A superb, detailed study of stability from the perspective of a modern-day ecologist is provided by S. L. Pimm (1991), *The Balance of Nature?,* University of Chicago Press, Chicago.

228–229 The adaptive landscape was first laid out in S. Wright (1932), *Proceedings of the Sixth International Congress of Genetics,* vol.1, pp.356–366. Influential geneticists, such as T. Dobzhansky, and paleontologists, such as G. G. Simpson, expanded on Wright's idea of adaptive peaks and valleys.

229–230 N. Eldredge (1989), *Macroevolutionary Dynamics,* McGraw-Hill, New York, p.34.

231 Richard Levins (1968), *Evolution in Changing Environments,* Princeton University Press, Princeton, N.J., pp.4, 10.

233 Crowley and North (1991), p.11.

234 G. C. Williams (1992), *Natural Selection,* Oxford University Press, New York, p.130.

236 The spectrum of environmental cycles: Crowley and North (1991), 140–141; B. Damnati (1993), *Journal of African Earth Sciences,* vol.16, pp.519–521; M. E. Schlesinger and N. Ramankutty (1994), *Nature,* vol.367, pp.723–726; L. A. Scuderi (1993), *Science,* vol.259, pp.1433–1436; G. Eglinton, et al. (1992), *Nature,* vol.356, pp.423–426; G. Bond, et al. (1992), *Nature,* vol.360, pp.245–249; Bond and R. Lotti (1995), *Science,* vol.267, pp.1005–1010; R. A. Kerr (1993), *Science,* vol.262, pp.1972–1973; and G. Plaut, et al. (1995), *Science,* vol.268, pp.710–713.

237 David Raup (1991), *Extinction,* Norton, New York.

240 Richard Dawkins (1987), *The Blind Watchmaker,* Norton, New York, p.181.

241 Matt Ridley (1993), *The Red Queen,* Macmillan, New York.

242–243 The thinking and research of evolutionary psychologists are clearly stated in J. H. Barkow, L. Cosmides, and J. Tooby, eds. (1993), *The Adapted Mind,* Oxford University Press, New York; and Tooby and Cosmides (1990), *Ethology and Sociobiology,* vol.11, pp.375–424.

247–248 K. V. Flannery, ed. (1986), *Guilá Naquitz,* Academic Press, Orlando.

249 Data that show rapid oscillations at the end of the Pleistocene: Englinton, et al. (1992), p.425.

VII: THE LITMUS TEST

259 An authoritative account of the Black Death is presented by Barbara W. Tuchman (1978), *A Distant Mirror: The Calamitous 14th Century,* Ballantine, New York, Chapter 5.

259–260 Lysenko and the effects of his precepts are recounted in Z. A. Medvedev (1969), *The Rise and Fall of T. D. Lysenko,* Columbia University Press, New York; D. Joravsky (1970), *The Lysenko Affair,* Harvard University Press, Cambridge, Mass.; D. Abbott, ed. "Lysenko, Trofim Denisovich" in *The Biographical Dictionary of Scientists: Biologists,* P. Bedrick Books, New York, pp.89–90.

260–263 The ideas and statements of what I have characterized as the Culture of Preservation are found in many sources, including McKibben (1989), and articles by Abrahamson, Beyea, Peters, and others in R. L. Wyman, ed. (1991), *Global Climate Change and Life on Earth,* Routledge, Chapman and Hall, New York. The values that give rise to this perspective are clearly presented in the volume edited by B. G. Norton (1986) and other books and articles already noted in Chapter II. The preservationist position is negatively characterized in the books by R. Bailey and D. L. Ray.

261 Aldo Leopold (1949), *A Sand County Almanac,* Oxford University Press, New York, pp.224–225.

263–266 The Culture of Growth perspective is less visible in the environmental literature but more apparent in the political and economic arenas. Two of the most daring presentations of this viewpoint are Ronald Bailey (1993), *Ecoscam: The False Prophets of Ecological Apocalypse,* St. Martin's Press, New York; and Dixy Lee Ray (1993), *Environmental Overkill,* Regnery Gateway, Washington, D.C.

266 A. Teramura's work is cited by D. J. Dudek (1991), "The Nexus of Agriculture, Environment, and the Economy Under Climate Change" in W. L. Wyman, ed. p.185; and R. Bailey (1993), pp.128–129.

268 R. Bailey (1993), p.67.

268 Evidence of a relatively stable environment over the past 8,000 years, based on the Greenland ice cores, is presented in references noted earlier for pp.167–168.

269 G. P. Marsh's statement, which is from his book *Man and Nature* (1864), is cited by C. Merchant (1989), p.241.

270 D. L. Ray (1993), p.208.

271 The statement about human effect on the world's primary productivity is based on P. M. Vitousek (1992), *Annual Review of Ecology and Systematics,* vol.23, p.8.

274 I first became aware of the litmus-test analogy in a 1990 presentation by Dr. Tom Simkin of the National Museum of Natural History, Washington, D.C.

276 R. Ardrey (1961); D. Morris (1969), *The Human Zoo,* Delta, New York; E. O. Wilson (1978), p.196.

ACKNOWLEDGMENTS

I thank the many scientists and authors on whose work I have relied, and hope that this personal synthesis of many fields has provided a fair representation. The solitude necessary to write was broken by several periods of fieldwork, leading to wonderful friendships in far-flung places, and the excitement and trials of scientific teams living together for weeks or months at a time. I especially thank my Olorgesailie research colleagues: Kay Behrensmeyer, Jennifer and Chip Clark, Alan Deino, Tom Jorstad, Bill Keyser, Bill Melson, Muteti Nume, Tom Plummer, Lassa Skinner, Nancy Sikes, our Kenya field crew, and the many students and other colleagues who have visited our field sites and have helped to make it a rich scientific and social experience. The National Museums of Kenya, particularly Meave Leakey and Mohamed Isahakia, deserve my heartfelt thanks for their encouragement and permission to conduct research in their country over the years. The presence and friendship of Mary Leakey have also inspired our field efforts. I thank my colleagues at the National Museum of Natural History, Smithsonian Institution, for permitting me the time off to write this book, and for putting up with the privacy I required to put ideas into words. Wendy Wiswall and Jennifer Clark generously assisted in the final manuscript production. My thanks also goes to Katinka Matson and John Brockman for their commitment to this book; Maria Guarneschelli for her initial involvement with it; Anne Freedgood, whose caring editorial eye saw the book to completion; and the staff of William Morrow and Company for treating me, without any basis, like an important author. Finally, I wish to express the deepest gratitude to my friends, particularly Lassa Skinner, and to my family for their support and love over the course of this project.

INDEX

INDEX

C

G

H

I

J

K

L

M

N

O

P

U

Y

Z